C000300454

THE ESSENCE OF

SOLID-STATE ELECTRONICS

THE ESSENCE OF ENGINEERING SERIES

Published titles
The Essence of Solid-State Electronics
The Essence of Electric Power Systems
The Essence of Measurement
The Essence of Engineering Thermodynamics

Forthcoming titles
The Essence of Analog Electronics
The Essence of Circuit Analysis
The Essence of Optoelectronics
The Essence of Microprocessor Engineering
The Essence of Communications
The Essence of Power Electronics

THE ESSENCE OF

SOLID-STATE ELECTRONICS

Linda Edwards-Shea
South Bank University

Prentice Hall
LONDON NEW YORK TORONTO SYDNEY TOKYO
SINGAPORE MADRID MEXICO CITY MUNICH

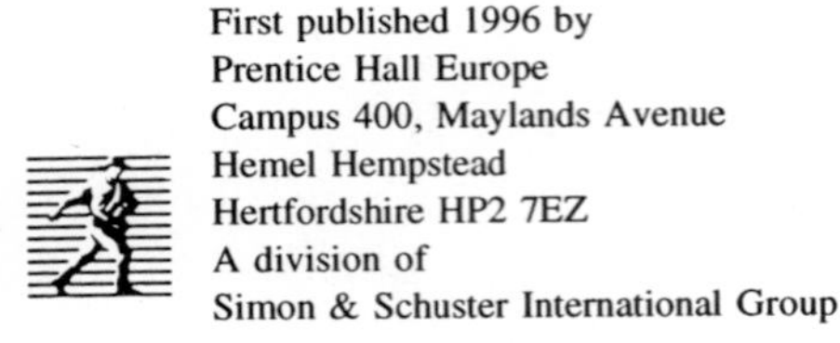

First published 1996 by
Prentice Hall Europe
Campus 400, Maylands Avenue
Hemel Hempstead
Hertfordshire HP2 7EZ
A division of
Simon & Schuster International Group

© Linda Edwards-Shea 1996

All rights reserved. No part of this publication may be reproduced, stored in a retrieval system, or transmitted, in any form, or by any means, electronic, mechanical, photocopying, recording or otherwise, without prior permission, in writing, from the publisher.
For permission within the United States of America contact Prentice Hall Inc., Englewood Cliffs, NJ 07632

Typeset in 10/12pt Times
by MHL Typesetting Ltd, Coventry

Printed and bound in Great Britain by
T.J. Press (Padstow) Ltd

Library of Congress Cataloging-in-Publication Data

Edwards-Shea, Linda.
The essence of solid-state electronics/Linda Edwards-Shea.
p. cm. — (The essence of engineering)
Includes bibliographical references and index.
ISBN 0-13-192097-9 (alk. paper)
1. Semiconductors. 2. Solid state electronics. I. Title.
II. Series.
TK7871.85.E414 1996
537.6′22–dc20 95-45890
CIP

British Library Cataloguing in Publication Data

A catalogue record for this book is available from the British Library

ISBN 0-13-192097-9

---

1 2 3 4 5 00 99 98 97 96

This book is dedicated to the memory of my parents,
Kath and John Shea.

Contents

Preface

This book is a concise introductory undergraduate text describing the essential solid-state electronics necessary for an understanding of the physics and operation of common electronic devices, such as p–n diodes and transistors. The book forms one semester's work and is suitable for science and engineering students studying Bachelor's-degree courses and BTEC higher-level courses. A tutor's guide containing answers to tutorial questions and suggestions for other topics of study has been written to accompany the book.

I started writing this book while lecturing in electronics to electrical engineering students at the City College of the City University of New York in Manhattan, New York City. At the time I was taking part in an exchange scheme between my home institution, South Bank University in London, England, and City College. I was struck by the similarities between the New York students and my London students. Many were mature, having worked for at least a year or two before continuing their education, and many had had a very patchy education, maybe leaving school without many qualifications or with arts qualifications and realizing later that they really wanted to return to education to fulfil their potential by becoming engineers.

While teaching the solid-state electronics part of the course I realized that many of the American students, like my British students, had difficulty finding suitable text books which were slanted to intelligent but often ignorant readers who needed a complete physical model which started from an introductory level. On looking around myself, I discovered that no such book existed. There are many books about solid-state electronics, but they concentrate either on a simple picture which I think is inadequate for undergraduate students, or they assume background skills like high-level arithmetic and the manipulation of sophisticated mathematical expressions, skills which many of my students had not had time to acquire.

Many of the students I have taught have not studied physics as a subject in its own right and so have been unable to acquire the vast, often unrecognized, culture that goes with physics—the skill and confidence needed to deal with very large or very small numbers, the ability to apply mathematics to real physical situations, the confidence to ask 'why?'. I decided such students needed a book which started with the Bohr model of the atom, which went on to the band theory of solids with an emphasis on semiconductors, and then on to semiconductor physics, the p–n junction, diodes, and transistors. Along the way they needed to grasp the concept of wave–particle duality, and they needed more general knowledge about electronic materials. Last, but by no means least, I wanted them to be able to buy a book which was affordable.

I therefore decided to write such a book myself. This book is designed to form the basis of an introductory solid-state electronics course for undergraduate students on engineering courses, or similar. It contains the essential information for a knowledge and understanding of the behaviour of common electronic materials and devices. The book is small and therefore not meant to be a reference book. It is designed for students to work through in chapter order, assessing themselves against the self-assessment questions as they proceed. Students who are part of a class for which this is a set book will be able to expand on some of the topics if their tutor uses the guide which accompanies the book.

Chapter 1 briefly describes the development of the Bohr model of the atom and introduces the principal quantum number for the quantization of electron energy.

Chapter 2 develops the quantum concept further by introducing the orbital, magnetic and spin quantum numbers and listing the mathematical rules which can be used to fill atoms with electrons. The periodic table of the elements is described. The electronic configurations of all the known elements are listed in Appendix A. I have included this large table because experience tells me that many students are peculiarly interested in the elements, even the very large, unstable ones which have been discovered and named recently, so I have included it for them.

The ideas expounded in the Bohr model are expanded in Chapter 3 to describe briefly the energy-band model for solids. The electron is treated as a subatomic sphere to do this. Solid materials are classified as conductors, insulators or semiconductors depending on their band structures. Elementary examination of cubic crystal structures and the use of Miller indices to denote crystal planes are included to enable readers to identify individual planes during the further study of semiconductors.

Semiconductors are described in Chapter 4 and the Fermi level is introduced to explain work function and electron affinity. The motion of holes through a semiconductor lattice is described in terms of the energy-band model. The role of doping is described using the pictorial covalent-bond model and the energy-band model. Equations for drift and diffusion are derived, and the chapter finishes by examining the Hall effect, photoconductivity and recombination.

The carrier-transport equations of Chapter 4 are used in Chapter 5 to build a simple model of the abrupt symmetrical p–n junction. Asymmetrical p–n junctions are examined briefly. Equations for depletion width and junction capacitance are derived. The use of junction capacitance to determine barrier height is described.

Chapter 6 examines semiconductors in more detail, beginning with an introduction to wave–particle duality. Schrödinger's wave equation is not covered—I make no apologies for this—and neither is the particle-in-a-box model. I think neither is *essential* for the great majority of engineers, even though they make an interesting and exciting model for further study. Chapter 6 also includes the use of Fermi–Dirac statistics for the determination of carrier concentration. The dependence of carrier concentration on temperature is explained, as are intrinsic Fermi level and intrinsic carrier concentration.

Chapter 7 extends the knowledge gained about the p–n junction by describing briefly the operation of common p–n junction devices: the rectifying diode, the zener diode, the photodetector, the solar cell, the light-emitting diode and the laser diode.

Chapter 8 continues the examination of common electronic devices by explaining the operation of field-effect transistors. Chapter 9 explains the operation of bipolar junction transistors.

Chapter 10 finishes the book with a brief review of some novel electronic materials, including diamond, superconductors and conducting polymers.

Self-assessment questions are scattered throughout the book, all with answers in the back. Each of the first nine chapters ends with a list of tutorial questions which take longer to answer. Solutions to these are in the *Guide for Tutors* which accompanies this book. A list of recommended texts for further reading is given at the end of each chapter.

Some of the chapters are more 'essential' than others: I would expect every scientist or engineer who uses electronic devices to be completely familiar with the contents of Chapters 1–5; all electronic engineers should also be familiar with Chapters 7–9. Anyone going on to further study in solid-state electronics should be completely familiar with the contents of the whole book. After working through this book readers should be well prepared to tackle more comprehensive books such as the excellent *Solid State Electronic Devices* by Ben Streetman.

Linda Edwards-Shea

Acknowledgements

I wish to thank all my current and former students at South Bank University and the City College of New York for their enthusiasm and constructive criticism, which have helped me write this book. Thanks must also go to Andrew Miller of *Chemistry and Industry* and Nigel Davis for information about the seaborgium affair, Anne Caroline and my sister Janet for their enthusiasm and perceptive comments, and Paul for making all those cups of tea and for reading and commenting on much of the manuscript.

How to use this book

I have written this book so that readers can start at Chapter 1 and *work* their way through to the end. Even when new readers are already familiar with the early material, I recommend that these readers still work through that material as a basis for revision.

At the start of each chapter there is a short list of aims and objectives, and at the end there is a summary of how these objectives have been reached. One of my main aims as an author is to provide the essential information for a knowledge and understanding of solid-state electronics, so these objectives may occasionally seem narrow in outlook. Suggested reading is listed at the end of each chapter so that readers can broaden their outlook.

Throughout the book there are self-assessment questions which should be answered as they are arrived at by the reader. I recommend that the reader keeps a notebook especially for this purpose. The answers should be recorded and checked against those in the back of the book. Usually the questions require short answers. Longer answers are usually required for the tutorial questions which come at the end of each chapter. The answers to these are included in the tutor's guide only. These tutorial questions are not just numerical problems — sometimes written answers are required.

Occasionally I have suggested some topics for research under the heading 'Research'. This research will be difficult for readers who do not have access to encyclopedias or a library, but I hope that carrying out the research will provide some breadth to the book, also some skill in finding and recording information. Again, I recommend the use of the notebook to record research findings and references.

I have assumed that students using this book will already have studied mathematics, including some calculus and rotational motion, or that they will be studying mathematics alongside their study of this book. In the early chapters the mathematics covered is not rigorous and I have stated several equations without saying where they've come from and without deriving them, but later in the book more mathematics is used to manipulate and derive equations.

I hope that by studying this book some readers will understand enough about solid-state electronics to be excited by the technology. I also hope they will go on to further study and practice of a fascinating subject.

A note about units

In this book I have endeavoured to impart skills as well as knowledge. To this end I have avoided the consistent use of SI units throughout the book, particularly for length and energy, the SI units of which are the metre (m) and the joule (J). I frequently use non-SI units of length such as cm, mm, μm, nm, pm and even Å. My favoured unit of energy is the electron volt eV, and the book is littered with these. Why have I disregarded the academic's love of SI units? Well, I love them too, but I also know that in industry the use of units is geared to the most appropriate choice for that industry. In the semiconductor industry, for example, the usual unit of length for device designers is the centimetre (cm) — probably because sample sizes are of the order of centimetres, certainly not metres. Likewise X-ray crystallographers use nanometres (nm) or angstrom units (Å) because they deal with lengths which are of the order of angstroms.

However, I do believe that an essential engineering skill is the ability to convert from SI to non-SI and back again. I have noticed in recent years that more and more students leaving school and entering my university find tremendous difficulty in dealing with non-SI units, and this has been reflected in conversations I have had with employers who complain bitterly that graduates these days seem incapable of converting from one system of units to another. Hence I have tried to tackle this problem in this book, especially in the early chapters, by giving readers some practice at converting between SI and non-SI.

Notation

Symbol	*Description*	*SI unit or most acceptable unit*
a	Lattice constant	m
$\mathbf{a}$	Basis vector	
A	Cross-sectional area of a sample	m^2
$\mathbf{b}$	Basis vector	
B	Base transport factor	
B_z	Manetic flux density in the x-direction	T
$\mathbf{c}$	Basis vector	
C_j	Junction capacitance	F
d	Orbital quantum number notation for $l = 2$	
D	Diffusion coefficient	$m^2\ s^{-1}$
D_n	Electron diffusion coefficient	$m^2\ s^{-1}$
D_p	Hole diffusion coefficient	$m^2\ s^{-1}$
$e\phi$	Work function	eV
$e\phi_m$	Metal work function	eV
$e\chi$	Electron affinity	eV
E	Energy	J, eV
E_a	Acceptor energy	eV
E_c	Conduction-band edge	eV
E_d	Donor energy	eV
E_f	Fermi level	eV
E_{fm}	Fermi level in a metal	eV
E_{fn}	Fermi level on the n-side of a p–n junction	eV
E_{fp}	Fermi level on the p-side of a p–n junction	eV
E_{fs}	Fermi level in a semiconductor	eV
E_g	Bandgap	eV
E_i	Intrinsic Fermi level	eV
E_j	Energy of an electron in the jth orbit	J, eV
E_{max}	Maximum energy of escaping electrons	J, eV
E_n	Energy of an electron in the nth orbit	J, eV
E_p	Energy of an electron in the pth orbit	J, eV

E_{ph}	Photon energy	eV
$E_{ph(max)}$	Maximum photon energy	eV
E_v	Valence-band edge	eV
$\mathcal{E}$	Electric field	V m^{-1}
$\mathcal{E}_H$	Hall field	V m^{-1}
$\mathcal{E}(x)$	Electric field in the x-direction	V m^{-1}
f	Orbital quantum number notation for $l = 3$	
f	Photon frequency	Hz
$f(E)$	Fermi–Dirac distribution function	
f_{pj}	Frequency of photon emitted by electron falling from p to j	Hz
F_{elec}	Electric force	N
F_{mag}	Magnetic force	N
g_m	FET transconductance	S
h_{FE}	h-parameter for common-emitter current gain	
hkl	Miller indices	
I	Current	A
I_B	Base current	A
I_C	Collector current	A
I_{Cn}	Electron component of the collector current	A
I_{Cp}	Hole component of the collector current	A
I_D	Drain current	A
I_{DSS}	Steady-state drain current	A
I_E	Emitter current	A
I_{En}	Electron component of the emitter current	A
I_{Ep}	Hole component of the emitter current	A
I_{max}	Maximum current obtained in solar-cell quadrant	A
I_o	Reverse saturation current	A
I_{sc}	Short-circuit current	A
I_x	Current in the x-direction	A
j	Electron orbit	
J	Current density	A m^{-2}
J_{drift}	Drift current density	A m^{-2}
$J_{n(diff.)}$	Electron-diffusion current density	A m^{-2}
$\mathrm{J}_{n(drift)}$	Electron-drift current density	A m^{-2}
$J_{p(diff.)}$	Hole-diffusion current density	A m^{-2}
$J_{p(drift)}$	Hole-drift current density	A m^{-2}
J_x	Current density in the x-direction	A m^{-2}
k	Wave vector	
l	Orbital quantum number	
L	Sample length	m
L_p	Hole diffusion length	m
m	Magnetic quantum number	
m	Mass	kg

m^*	Effective mass	kg
m_n^*	Electron effective mass	kg
m_p^*	Hole effective mass	kg
m_r	Relativistic mass	kg
n	Electron orbit, principal quantum number	
n	Electron concentration	m^{-3}
n_i	Electron intrinsic carrier concentration	m^{-3}
n_n	Electron concentration on n-side of p–n junction	m^{-3}
n_o	Electron concentration at thermal equilibrium	m^{-3}
n_p	Electron concentration on p-side of p–n junction	m^{-3}
$\frac{dn(x)}{dx}$	Electron concentration gradient in the x-direction	m^{-4} or cm^{-4}
N_a	Acceptor concentration	m^{-3}
$N_{aEmitter}$	Acceptor density in the emitter of a pnp BJT	m^{-3}
N_c	Effective density of states in the conduction band	m^{-3}
N_d	Donor concentration	m^{-3}
N_{dBase}	Donor density in the base of a pnp BJT	m^{-3}
$N(E)$	Density of energy states at energy E	m^{-3}
N_v	Effective density of states in the valence band	m^{-3}
p	Orbital quantum number notation for $l = 1$	
p	Electron orbit	
p	Hole concentration	m^{-3}
p_i	Hole intrinsic carrier concentration	m^{-3}
p_n	Hole concentration on the n-side of a p–n junction	m^{-3}
p_o	Hole concentration at thermal equilibrium	m^{-3}
p_p	Hole concentration on the p-side of a p–n junction	m^{-3}
$\frac{dp(x)}{dx}$	Hole concentration gradient in the x-direction	m^{-4}
P_{max}	Maximum output power from a solar cell	W
P_{solar}	Incident power reaching a solar cell	W
Q	Total charge	C
Q_n	Total charge on the n-side of a p–n junction	C
Q_p	Total charge on the p-side of a p–n junction	C
r_j	Radius of the jth electron orbit	m
r_n	Radius of the nth electron orbit	m
r_p	Radius of the pth electron orbit	m
R	Resistance	Ω
R_B	Base resistance	Ω
R_C	Collector resistance	Ω
$R_{channel}$	FET channel resistance	Ω
$R_{forward}$	Resistance in forward bias	Ω
R_H	Hall coefficient	$m^3\ C^{-1}$
R_L	Load resistance	Ω
$R_{reverse}$	Resistance in reverse bias	Ω
s	Orbital quantum number notation for $l = 0$	

s	Spin quantum number	
t	Sample thickness	m
T	Absolute temperature	K
v	Velocity	m s^{-1}
v_{drift}	Drift velocity	m s^{-1}
v_{in}	A.C. input voltage	V
v_{out}	A.C. output voltage	V
$v_{thermal}$	Thermal velocity of an electron	m s^{-1}
v_x	Velocity in the x-direction	m s^{-1}
$V_{applied}$	Applied voltage	V
V_{br}	Breakdown voltage	V
V_{BB}	Base bias voltage	V
V_{CC}	Collector bias voltage	V
V_{DS}	Voltage between drain and source	V
$V_{DS(p)}$	Pinch-off value of V_{DS} for a given value of V_{GS}	V
V_f	Forward bias voltage	V
V_{GD}	Voltage between gate and drain	V
V_{GS}	Voltage between gate and source	V
$V_{GS(off)}$	Cur-off voltage in a FET	V
V_H	Hall voltage	V
V_{max}	Maximum voltage obtained in solar-cell quadrant	V
V_n	Voltage on the n-side of the depletion width	V
V_o	Contact potential	V
V_{oc}	Open-circuit voltage	V
V_p	Voltage on the p-side of the depletion width	V
V_r	Reverse bias voltage	V
V_T	Threshold voltage	V
$V_{turn-on}$	Turn-on voltage, threshold voltage	V
$\frac{-dV(x)}{dx}$	Voltage (potential) gradient in the x-direction	V m^{-1}
V_Z	Zener voltage	V
w	Sample width	m
W	Depletion width	m
x	x-direction	
$+x_n$	Penetration of the depletion width into the n-side of a p–n junction	m
$-x_p$	Penetration of the depletion width into the p-side of a p–n junction	m
y	y-direction	
z	z-direction	
Z	Atomic number	
α	Common-base current gain	
β	Common-emitter current gain	
γ	Emitter injection efficiency	

ϵ_r	Relative permittivity, dielectric constant	
ϵ_s	Semiconductor permittivity	F m^{-1}
η_s	Solar efficiency	
λ	Wavelength	m
λ_{dBr}	de Broglie wavelength	m
λ_{pj}	Wavelength of photon emitted between oribts p and j	m
μ	Drift mobility	m^2 V^{-1} s^{-1}
μ_n	Electron drift mobility	m^2 V^{-1} s^{-1}
μ_p	Hole drift mobility	m^2 V^{-1} s^{-1}
ρ	Resistivity	Ω m
σ	Conductivity	Ω^{-1} m^{-1}
σ_n	Electron conductivity	Ω^{-1} m^{-1}
σ_p	Hole conductivity	Ω^{-1} m^{-1}
τ	Mean free time between collisions or carrier lifetime	s
τ_n	Electron lifetime	s
τ_p	Hole lifetime	s
ϕ	Work function	V
ϕ_m	Metal work function	V
$\Phi_n(x)$	Electron flux density in the x-direction	m^{-2} s^{-1}
$\Phi_p(x)$	Hole flux density in the x-direction	m^{-2} s^{-1}
χ	Electron affinity	V

Constants used in this book

Quantity	*Symbol*	*Value and unit*
Velocity of light *in vacuo*	c	2.998×10^{8} m s^{-1}
Magnitude of electronic charge	e	1.602×10^{-19} C
Planck's constant	h	6.626×10^{-34} J s
Reduced form of Planck's constant	$\hbar$	1.055×10^{-34} J s
Boltzmann's constant	k	1.381×10^{-23} J K^{-1} or 8.620×10^{-5} eV k^{-1}
Rest mass of the electron	m_o	9.109×10^{-31} kg
Avogadro's number	N_A	6.022×10^{23} mol^{-1}
Rydberg constant	R	1.097×10^{7} m^{-1}
Permittivity of free space	ϵ_o	8.854×10^{-12} F m^{-1} or 8.854×10^{-14} F cm^{-1}
Approximate thermal energy of an electron at room temperature	kT/e	0.025 eV or 1/40 eV

The Greek alphabet

Name	*Lower case*	*Upper case*
Alpha	α	A
Beta	β	B
Gamma	γ	Γ
Delta	δ	Δ
Epsilon	ϵ	E
Zeta	ζ	Z
Eta	η	H
Theta	θ	Θ
Iota	ι	I
Kappa	κ	K
Lambda	λ	Λ
Mu	μ	M
Nu	ν	N
Xi	ξ	Ξ
Omicron	o	O
Pi	π	Π
Rho	ρ	P
Sigma	σ	Σ
Tau	τ	T
Upsilon	υ	Υ
Phi	ϕ	Φ
Chi	χ	X
Psi	ψ	Ψ
Omega	ω	Ω

Unit prefixes

Prefix	*Name*	*Multiplier*
E	exa	10^{18}
P	peta	10^{15}
T	tera	10^{12}
G	giga	10^{9}
M	mega	10^{6}
k	kilo	10^{3}
h	hecto	10^{2}
da	deca	10^{1}
d	deci	10^{-1}
c	centi	10^{-2}
m	milli	10^{-3}
μ	micro	10^{-6}
n	nano	10^{-9}
p	pico	10^{-12}
f	femto	10^{-15}
a	atto	10^{-18}

CHAPTER 1

Electrons in atoms

Aims and objectives

The aim of this chapter is to explain the behaviour of atoms by describing a model of how their electrons behave. Bohr proposed a model of the atom which was so accurate that the behaviour of the simplest atom, hydrogen, could be predicted and explained. Modern theories of electron behaviour, such as quantum theory, were then developed from Bohr's model. In this chapter you will be introduced to the ideas and models which led to quantum theory. The results of some experiments, such as the photoelectric effect and the glow discharge from hydrogen, will be explained. You'll be able to use the equations describing electrons in atoms to determine some of the properties of the atoms, such as size, electron energy and ionization energy.

1.1 Atomic models

A set of ideas which leads to a description of something is called a model. Models are important because they can be used to predict experimental results. We call those ideas which lead to the description of the atom an atomic model.

Everything is made up of atoms. In 1911, Ernest Rutherford (1871–1937) suggested that an atom consisted of a positively charged nucleus with negatively charged electrons orbiting the nucleus at constant speed (Figure 1.1). He was awarded the Nobel Prize for Chemistry in 1908 for his discovery of the atomic nucleus.

Note that, according to Rutherford's model, most of the atom is made up of empty space, such that almost all of the atom's mass is concentrated in the nucleus. If we imagine the nucleus magnified to the size of a hockey ball, the outermost part of the atom would be miles away. This was a major departure from earlier models, which proposed that the electrons were stuck into the spherical surface of the nucleus, rather like a plum pudding.

Niels Bohr (1885–1962), a Danish physicist, made a great contribution to our understanding of the world by devising an atomic model which explains what matter is and why it has particular properties. In fact, his model explains the behaviour of hydrogen only, but it led to other models which describe the behaviour of all the known elements. Bohr derived his model from Rutherford's while working as one of Rutherford's research students in Cambridge, England. This chapter will explain how

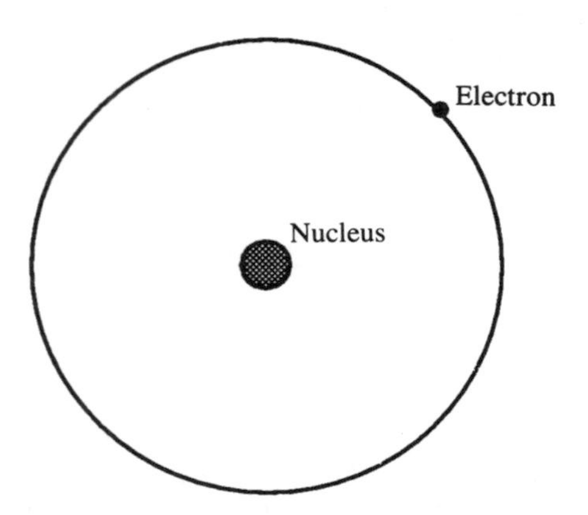

Figure 1.1 *Rutherford's simple atomic model.*

Bohr's model was developed and how it predicted the observed results of experiments on hydrogen gas. Niels Bohr was awarded the Nobel Prize for Physics in 1922.

1.2 Glow discharges and their line spectra

Before we examine Bohr's model we should consider some experiments that were done just before Bohr devised his model. If hydrogen gas is sealed in an evacuated glass tube and a high voltage is applied across it, the gas glows. This is called a glow discharge and the tube in which it occurs is called a glow-discharge tube (Figure 1.2).

The light, or electromagnetic radiation, making up the glow can be analyzed by passing it through a piece of optical equipment called a spectrograph. The spectrograph splits the light up into its constituent wavelengths such that all the light of a certain wavelength hits a screen (a piece of photographic film) in the same place. Hence an image of the slit illuminated by the light leaving the spectrograph is obtained (Figure 1.3). The result is a series of sharply defined, discrete lines called a line spectrum. Each line is an image of the entrance slit in the spectrograph. Each and every line represents a single wavelength of light emitted by the hydrogen discharge. Several series of these lines are obtained for hydrogen and, because they're so important, they're named after the scientists who first reported them (Lyman, Balmer, Paschen).

The line spectra for hydrogen were very important to Niels Bohr because they gave him vital clues about the behaviour of hydrogen gas, which he happened to be studying himself. Somebody else, Rydberg, had also been studying hydrogen. Rydberg had discovered an empirical relationship which described the line spectra. This means that he had found that he could fit numbers into an equation which predicted the wavelengths represented by the lines, and the wavelengths obtained using the equation matched the wavelengths measured on the film. This relationship is:

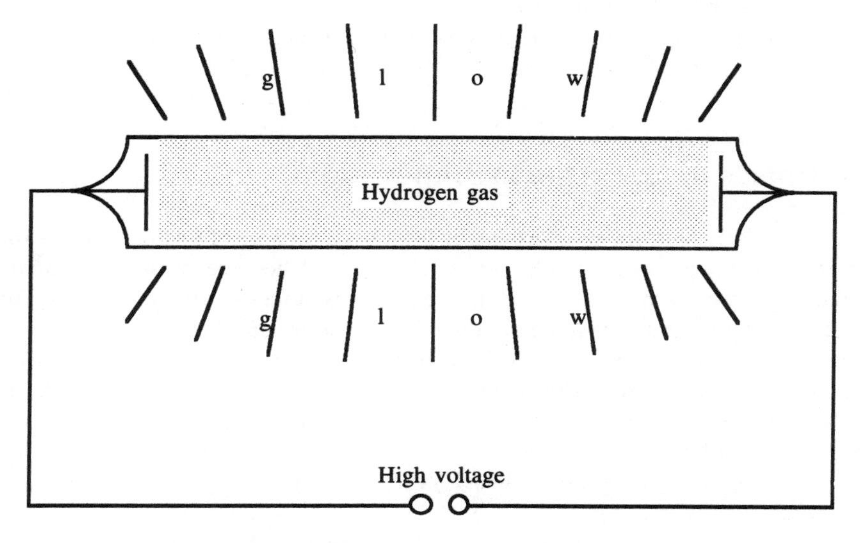

Figure 1.2 *A glow-discharge tube containing hydrogen.*

$$\frac{1}{\lambda_{pj}} = R\left(\frac{1}{j^2} - \frac{1}{p^2}\right) \tag{1.1}$$

where $j < p$, and j and p are integers. R is the Rydberg constant and λ_{pj} is the wavelength of light in the glow which depends on the values p and j. R is a fundamental constant — you'll become familiar with such constants because they appear in many equations describing electronic devices.

Rydberg constant $R = 1.097 \times 10^7 \text{ m}^{-1}$

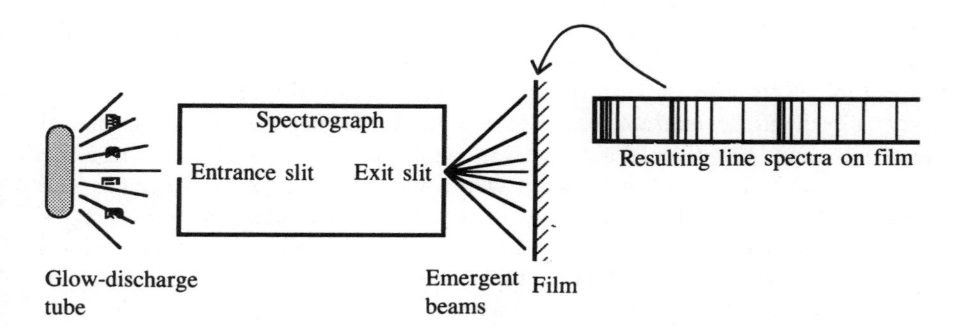

Figure 1.3 *Schematic diagram of a line spectrograph.*

SELF-ASSESSMENT QUESTION 1.1

What is the wavelength of the photon emitted when $p = 4$ and $j = 2$? *Hint*: Use equation (1.1).

1.3 Bohr's model of the hydrogen atom

Bohr knew the results of the work done on the glow discharge of hydrogen, and so his aim was to devise a model for hydrogen which predicted the same results. Once Bohr's predictions matched the observed results he knew his model was basically right. This is the way in which scientists learn about the world around them.

Hydrogen is the simplest atom. It also happens to be the smallest atom and normally contains only one electron. The quantity of electric charge on an electron is a basic, fundamental quantity of electricity. This numerical value (1.6×10^{-19} coulombs) is given the symbol e (or, sometimes, q). Note that it doesn't really matter which symbol you use (e or q) but you should say which you're going to use *before* you use it. So, in this book,

$$\text{electronic charge } e = 1.602 \times 10^{-19}\ \text{C}$$

In 1913, when Niels Bohr was 28 years old, he was a postgraduate student working in Rutherford's laboratory. Like other workers at the time, Bohr chose to model hydrogen rather than any other element because it was known that hydrogen had a simple atomic structure. By this time it was already established, by Rutherford, that the hydrogen atom consisted of a positively charged nucleus with one electron circling it at a constant velocity. He also knew that the hydrogen nucleus had a positive charge $+e$ and that the electron had a negative charge $-e$. He also realized that the electron must be bound to the nucleus by a force — if it were not, there'd be nothing to keep the electron with the nucleus and the electron would just wander off. This force is electrostatic in nature and exists because of the attraction between the positive nucleus and the negative electron.

Bohr found an inconsistency in Rutherford's version of the atom. Rutherford had suggested that the electron orbited the nucleus in a circular orbit at constant velocity v. This implied that the electron would lose energy as it orbited the nucleus because of the energy it would use in travelling around the nucleus. If this were the case, one would expect the electron to spiral in towards the nucleus as it proceeded in its orbit so eventually the atom would collapse. We know this doesn't happen because, if it did, all materials would just disintegrate — this is contrary to our experience. It would also mean that the hydrogen glow discharge would not radiate discrete wavelengths, and we know that wouldn't be right because there is the evidence of discrete wavelengths on the line spectra. If the electron radiated energy as it spiralled in towards the nucleus, we would expect to see a continuous blur representing a continuous range of wavelengths on the spectrograph film, and not the discrete lines that were observed. Bohr realized this and so he knew that Rutherford's ideas were too simple, so he developed his own model to describe the behaviour of hydrogen atoms. He set out to reconcile the idea of an orbiting electron with the experimental observations from the glow-discharge experiments. To achieve this, Bohr made three postulates.

1.3.1 Bohr's first postulate

Bohr liked the idea of the orbiting electron. He postulated that, since the orbiting electron couldn't possibly move in towards the nucleus (because he knew that hydrogen atoms were stable and didn't radiate a continuous range of energies and didn't collapse),

> there are certain orbits in which the electron is stable and does not radiate.

Remember that Bohr was developing a model to describe hydrogen atoms such that the model would accurately predict experimental observations (this is the purpose of a model). So it was perfectly acceptable for Bohr to propose this postulate — he could have proposed anything he liked. So let's consider an atom in which electrons move in stable orbits, such as postulated by Bohr (Figure 1.4). Call the radii of these orbits r_1, r_2, etc., where r_p is the radius of the pth orbit.

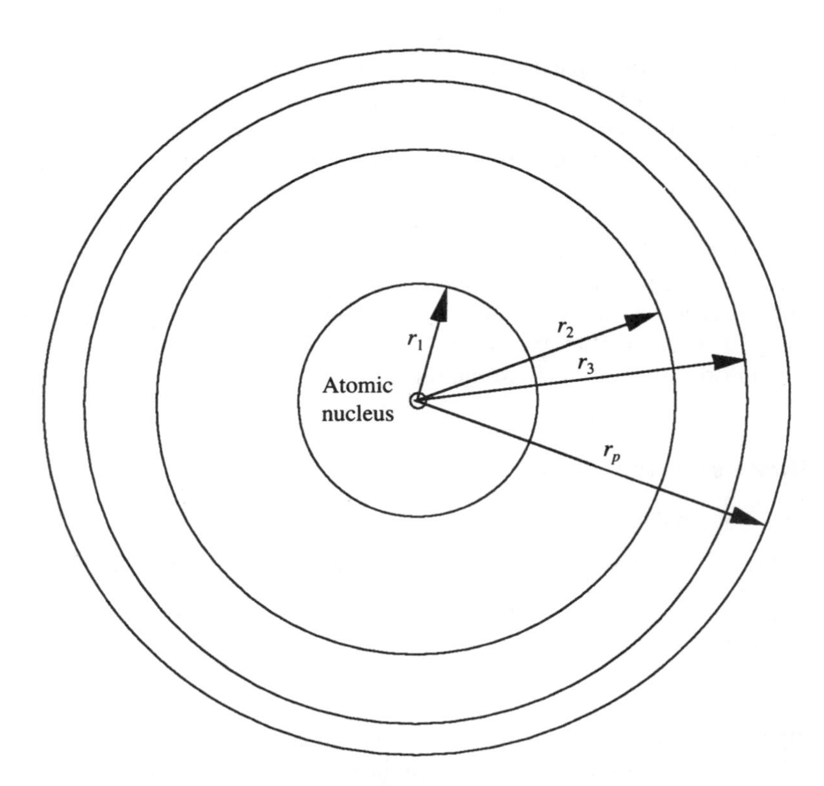

Figure 1.4 *Stable orbits of radius r around the atomic nucleus*

We can write an equation for the energy of an electron moving in the pth orbit:

$$E_p = -\frac{e^2}{8\pi\epsilon_o r_p} \text{ joules} \tag{1.2}$$

where ϵ_o is a constant called the permittivity of free space and E_p is the energy of an electron in the pth orbit. Remember e is the electronic charge, 1.6×10^{-19} C.

$$\text{Permittivity of free space } \epsilon_o = 8.854 \times 10^{-12} \text{ Fm}^{-1}$$

Equation (1.2) shows that, as the electron moves further away from the nucleus (i.e. as r_p gets larger) the right-hand side of the equation becomes smaller, i.e. less negative. Hence the electron's energy will have to become more positive (i.e. less negative).

SELF-ASSESSMENT QUESTION 1.2

According to Bohr's first postulate, if an orbiting electron in a hydrogen atom has energy equal to -2.18×10^{-18} J, how far out must it be from the atomic nucleus? Give your answer to two significant figures. *Hint*: Use equation (1.2).

Bohr's first postulate meant that an electron could be found only in one of the stable orbits because Bohr had *ruled* that electrons couldn't exist anywhere else. This led to the possibility of an electron being able to move from one orbit to another, either by jumping from a lower to a higher orbit or by falling from a higher to a lower orbit — so long as the electron didn't end up *between* orbits, these jumps and falls would be permissible. It became clear to Bohr that the high voltage needed to make hydrogen glow in a glow-discharge tube could be responsible for putting energy into the hydrogen atoms themselves, and this input of energy could give each electron itself more energy, such that it would jump from a lower orbit to a higher orbit. From equation (1.2) we know that more energy is required when the electron is in a higher orbit (i.e. an orbit further out from the nucleus).

1.3.2 Bohr's second postulate

Bohr then had another idea (his second postulate):

when the electron falls from orbit p to orbit j, the energy which it loses, $E_p - E_j$, is radiated as a quantum of light (Figure 1.5).

The equation

$$E_{\text{ph}} = hf \tag{1.3}$$

is an important fundamental equation describing a quantum of light (a photon). The energy E_{ph} contained in a photon is the product of the photon frequency f and an important constant called Planck's constant, h.

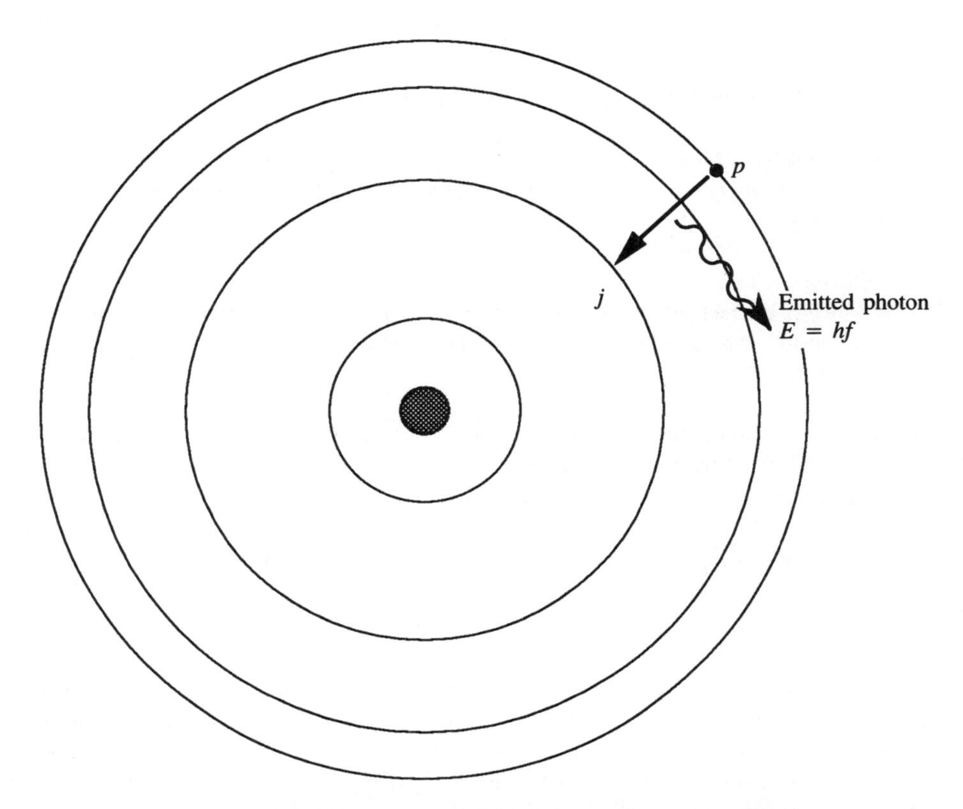

Figure 1.5 *A photon is emitted when the electron falls from orbit p to orbit j.*

$$\text{Planck's constant } h = 6.626 \times 10^{-34} \text{ J s}$$

Note that the photon frequency is not the frequency at which photons are emitted. It's the frequency of the photon itself. Figure 1.6 shows how a photon can be represented as a 'packet' of light, having a sinusoidal wave structure and a finite length. Being a wave, a photon has a wavelength λ and a frequency f. Also, the photon will travel at the speed of light, c, another important constant:

$$\text{speed of light in vacuo } c = 2.998 \times 10^{8} \text{ m s}^{-1}$$

The frequency f can be expressed as

$$f = \frac{c}{\lambda} \tag{1.4}$$

hence photon energy E_{ph} can be written

$$E_{\text{ph}} = \frac{hc}{\lambda} \tag{1.5}$$

by substituting equation (1.4) into equation (1.3).

SELF-ASSESSMENT QUESTION 1.3

How much energy is contained in a photon of wavelength 1 μm? What is its frequency?

RESEARCH 1.1

Before you tackle the tutorial questions at the end of Chapter 1, find out the range of wavelengths that is visible to the human eye. Make a note of the references you've consulted.

We now know enough to write an equation representing the energy lost by the electron as it falls from orbit p to orbit j:

$$E_p - E_j = hf_{pj} \tag{1.6}$$

Using equations (1.2) and (1.3), equation (1.6) can be re-written

$$\frac{1}{\lambda_{pj}} = \frac{e^2}{8\pi\epsilon_o hc}\left(\frac{1}{r_j} - \frac{1}{r_p}\right) \tag{1.7}$$

So we now have a useful equation for the wavelength of light which should, if Bohr's model is correct, be emitted by hydrogen atoms. This equation shows that the wavelength of an emitted photon is related to the orbits p and j between which the electron falls. Bohr already knew what these wavelengths were from the line spectra (Figure 1.3). If equation (1.7) produced the same wavelengths for the glow in hydrogen as were measured from hydrogen's line spectra, his model would be

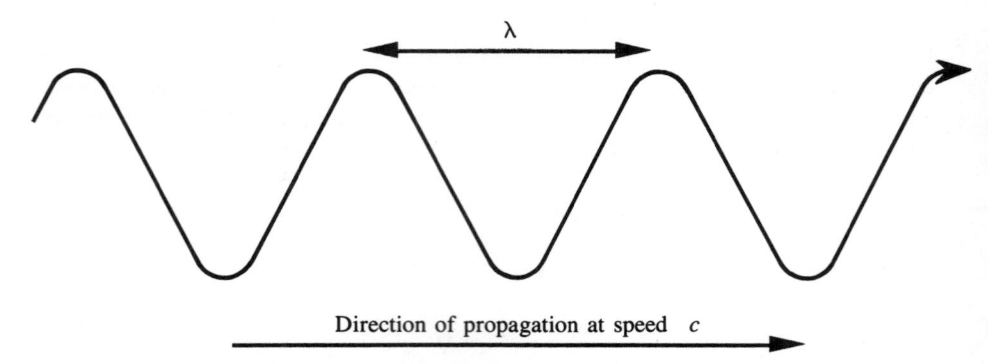

Figure 1.6 *Schematic representation of a photon.*

accurate. You may have noticed that equation (1.7) is similar in form to equation (1.1). Bohr found a way of making equations (1.1) and (1.7) coincide exactly: he proposed his third postulate.

1.3.3 *Bohr's third postulate*

This postulate took the form of an equation:

$$m_o v r_n = \frac{nh}{2\pi} \qquad (1.8)$$

where m_o is the mass of the electron (called the rest mass), v is the velocity of the electron as it travels around the orbit, n is an integer used to label individual orbits called the principal quantum number, and r_n is the radius of the nth orbit. The mass of the electron is another constant which you'll become familiar with:

$$\text{rest mass of the electron } m_o = 9.109 \times 10^{-31}\,\text{kg}$$

For now, all you need to know is that the mass of an electron varies depending on the medium it's moving through. Rest mass is the mass of a hypothetical electron which is stationary, i.e. not moving from atom to atom.

By making his third postulate, Bohr found that he could rewrite the empirical relationship equation (1.1) as:

$$\frac{1}{\lambda_{pj}} = \frac{e^4 m_o}{8 e_o^2 h^3 c}\left(\frac{1}{j^2} - \frac{1}{p^2}\right) \qquad (1.9)$$

Hence

$$\frac{e^4 m_o}{8 \epsilon_o^2 h^3 c}$$

must be the Rydberg constant R! This was a very exciting discovery because it confirmed that Bohr's model really did describe what was happening in hydrogen atoms. This was the first time an atomic model had been devised which accurately explained experimental observations. It was a turning point in the scientific progress of the twentieth century.

Excitation potential

Excitation occurs when the orbiting electron receives an amount of energy such that the electron is excited from a lower into a higher orbit around the atom — the energy is not sufficient to tear the electron away completely from the atom.

Electrons for $n = 1$ are tightly bound to the nucleus because of the strong electrostatic forces between the nucleus and the electrons which are close to it. As n increases, however, the orbit radius increases and the electrons become less tightly bound. Outer electrons are therefore more easily excited than inner electrons. After an electron is excited, it will return to its unexcited state (called the ground state) by the emission of a photon of energy E_{ph} and wavelength λ. The excited electron spends only a short time in the excited state, about 10 ns or so, before returning to its ground state.

SELF-ASSESSMENT QUESTION 1.4

Check for yourself that $e^4 m_o / 8\epsilon_o^2 h^3 c$ does equal the Rydberg constant.

1.4 Ionization potential

So far we have considered a neutral hydrogen atom in which an electron orbits the nucleus, and we have described that atom using Bohr's model. Ionization potential is the energy which has to be applied to the atom to strip the electron away from the atom (Figure 1.7), leaving the positively charged nucleus (a positive ion).

We've seen that Bohr's model assumed that the single electron in a hydrogen atom could exist in orbits with certain radii, given by Bohr's third postulate, equation (1.8):

$$m_o v r_n = \frac{nh}{2\pi}$$

such that

$$r_n = \frac{nh}{2\pi m_o v} \tag{1.10}$$

We've also seen that these orbits can have total energy given by equation (1.2):

$$E_n = -\frac{e^2}{8\pi\epsilon_o r_n}$$

This equation can be rewritten

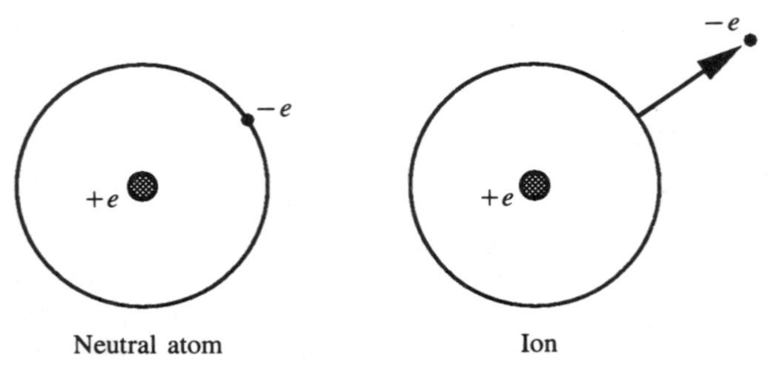

Figure 1.7 Ionization of a hydrogen atom.

$$E_n = -\frac{1}{n^2}\frac{m_o e^4}{8\epsilon_o^2 h^2} \tag{1.11}$$

This equation is very useful because it provides a practical way of calculating E_n, the energy of an electron in the nth orbit. We know the constants m_o, e, ϵ_o *and* h to a high degree of accuracy. We also know n — Bohr's model tells us it's an integer with a lowest possible value of one. The only unknown is E_n, which is easily calculated.

The unit of energy in equation (1.11) is the joule (symbol J). If you were to calculate E_n in joules you'd find that E_n would be very small — of the order of 10^{-18} J. Electronic engineers use electron energies frequently as part of calculations or as a means of specifying devices, so it's desirable to have a unit which is more manageable than the joule. Consequently we use the electron volt (symbol eV).

RESEARCH 1.2 18-2-97

Find the definition of the electron volt and write it in your notebook, noting the reference.

Converting from joules to electron volts is easy. To convert from joules to electron volts, divide the energy in joules by the electronic charge e.

$$\text{energy in eV} = \frac{\text{energy in J}}{\text{electronic charge}}$$

Conversely, to convert from electron volts to joules, multiply the energy in electron volts by e.

So, how do we go about evaluating the ionization energy of hydrogen? Hydrogen has one electron which will be in the first orbit in an unexcited atom. If we calculate the energy of that electron it'll be equal to the amount of energy needed to strip the electron away from the influence of the nucleus. That is, the ionization energy of hydrogen is equal to the energy of the electron in the orbit $n = 1$.

SELF-ASSESSMENT QUESTION 1.5

Calculate the ionization energy of hydrogen giving your answer in joules and in electron volts. *Hint*: Find E_n for $n = 1$.

Tables are published showing ionization energies for different atoms. If you look at these tables you'll see that the ionization energy of hydrogen is relatively high. This is because all other atoms are larger and have more electrons, so the outermost electron in these atoms isn't so tightly bound to the nucleus by electrostatic attraction, therefore less energy is required to remove the electron from the atom.

1.5 The photoelectric effect

Einstein (1879–1955) interpreted the results of an experiment which clearly demonstrated the discrete nature (quantization) of light. He was awarded the Nobel Prize for Physics for this work in 1946.

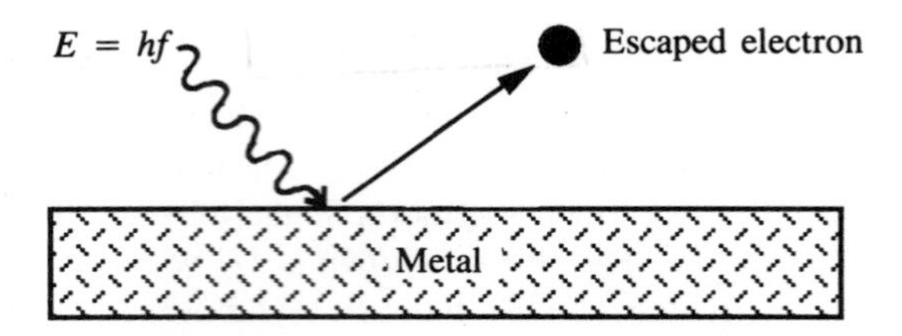

Figure 1.8 *Schematic diagram illustrating the photoelectric effect.*

1.5.1 A practical demonstration of quantum theory

The experiment involves a slab of metal which sits in an evacuated glass tube. The electrons in the metal absorb energy from light which is used to illuminate it. Einstein found a relationship between the amount of energy absorbed by the metal and the frequency of the light. Suppose monochromatic light (light of only one wavelength) is incident on the surface of a metal plate in a vacuum (Figure 1.8). The electrons in the metal absorb energy from the light and some of the electrons receive enough energy to be ejected from the metal surface into the vacuum. This phenomenon is known as the photoelectric effect.

If the energy of the escaping electrons is measured, a graph can be drawn of the maximum energy E_{max} versus the frequency f of the incident light (Figure 1.9). The slope of the resulting straight line is equal to Planck's constant h.

The equation describing the plotted line is

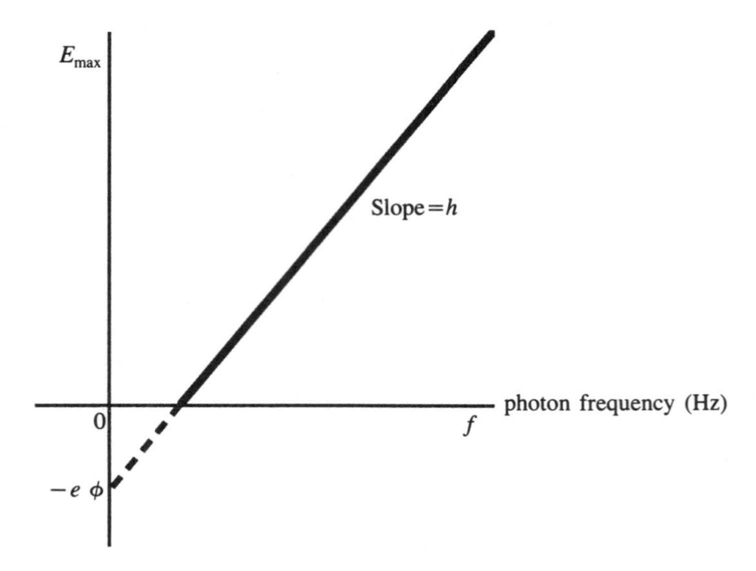

Figure 1.9 *Plot of E_{max} versus f to determine h.*

$$E_{max} = hf - e\phi \tag{1.12}$$

where h = Planck's constant, f = frequency of incident light ($f = c/\lambda$), e = electronic charge, ϕ = a characteristic of the metal in volts.

When ϕ is multiplied by the electronic charge e, an energy is obtained which represents the minimum energy required for an electron to escape from the metal. This energy $e\phi$ is called the work function of the metal. These results indicate that the electrons gain an amount of energy hf from the incident light and lose an amount of energy $e\phi$ in escaping from the metal surface.

The experiment demonstrates that light energy is contained in discrete units rather than in a continuous distribution of energies. Hence it shows that $E_{ph} = hf$ joules and demonstrates the quantized nature of light. A quantum of light is called a photon and can be considered either as a wave or a particle.

The photoelectric effect was therefore important because it led the way to the development of quantum theory and modern physics.

RESEARCH 1.3

Find work-function values for some common metals such as copper, gold and aluminium. Write your findings in your notebook. Also list the references so that, if required, you can find the same material again.

1.6 **Summary**

Bohr had arrived at the emission spectrum of hydrogen by using the Rutherford model as a starting point and by making three postulates:

1 There are certain orbits in which the electron is stable and does not radiate.
2 When the electron falls from a higher to a lower orbit the energy it loses is radiated as a quantum of light (a photon).
3 $m_o v r_n = nh/2\pi$.

An electron moves around the atomic nucleus in a stable orbit and can be excited into a higher stable orbit if energy is absorbed by the atom (such as the electrical energy arising from the high voltage connected across a glow-discharge tube). Light is then emitted when the electron falls from one stable orbit to a lower stable orbit, described by equation (1.7). There is a natural tendency for an excited electron to fall back towards its original orbit after being excited, shown by the evidence of the glow in the discharge tube. Also, only certain stable orbits exist, the radii of which are given by rearranging equation (1.8), Bohr's third postulate:

$$m_o v r_n = \frac{nh}{2\pi}$$

to give

$$r_n = \frac{nh}{2\pi m_o v}$$

Thus we've seen how Bohr's model successfully explains the observed experimental results for the light emitted from hydrogen atoms.

The photoelectric experiment demonstrated that light energy is contained in discrete (i.e. quantized) units called photons rather than in a continuous distribution of energies.

1.7 Tutorial questions

1.1 Calculate the energies (in joules and in electron volts) and radii of the second, third, fourth and fifth orbits of a hydrogen atom. Represent them on an energy-level diagram showing n, orbit radius r_n (in angstrom units) and electron energy E_n (in electron volts). *Hint*: Use graph paper to show your energy-level diagram to scale, by drawing a vertical axis representing E_n and a horizontal line to represent each energy level. Mark n and r_n on each energy level you draw. Alternatively, use the vertical axis to represent r_n and mark on values of n and E_n.

1.2 What is the wavelength of the photon that would be emitted by an electron falling from $n=5$ to $n=3$ in a hydrogen atom? If there were enough of them, would such photons be visible to the human eye?

1.3 If a photon emitted from hydrogen in a glow-discharge tube has a wavelength of 0.485 μm and an electron caused the emission by falling to an orbit which is 2.12 Å from the nucleus, how far is the orbit from which it fell from the nucleus? Draw a sketch to illustrate the problem.

1.8 Suggested further reading

Anderson, J. C., Leaver, K. D., Rawlings, R. D. and Alexander, J. M., *Materials Science*, 3rd edn, Van Nostrand Reinhold, 1985, pp. 50–52.

CHAPTER 2

Quantum numbers and the Periodic Table of the Elements

Aims and objectives

We've looked at the difference between excitation potential and ionization potential, and we've talked in terms of 'allowed' orbits or energy levels which orbiting electrons can occupy. We've also used the symbol n where $n = 1, 2, 3, \ldots$ to show how these orbits have discrete values of energy E_n, depending on this important number n. In fact, n is called the principal quantum number and it turns out that we've already arrived at quantum theory, which we'll use to explain the electronic behaviour of atoms.

People talk a lot of nonsense about quantum theory, usually saying it's very difficult, grandiose, esoteric, abstract, etc., etc. The fact is, quantum theory can be studied at quite a simple level without losing sight of its importance and underlying meaning, and without distorting it so much that you have to forget it all later on if you need to study more deeply. So welcome to quantum theory — the jewel of the twentieth century's scientific discoveries, and the foundation of the modern electronics industry.

2.1 Energy levels

Figure 2.1 shows the energy levels of the hydrogen atom. These are obtained using Bohr's model. The diagram is really an energy scale, because electron energy is measured vertically whereas the horizontal axis doesn't really measure anything. The values of E_n can be found using

$$E_n = -13.6/n^2 \text{ eV} \tag{2.1}$$

where n is the principal quantum number. Note that equation (2.1) is the same as equation (1.11) with the constants evaluated. You can see from Figure 2.1 that the lowest value of n is 1, representing the energy level (or orbit) nearest the nucleus.

E_n values are represented in Figure 2.1 by the horizontal lines across the scale. For example, the $n = 1$ line appears at $E_n = -13.6$ eV because hydrogen's electron has that energy when the electron is in the first Bohr orbit (i.e. $n = 1$). The $n = 2$ energy level occurs at $-13.6/2^2$ eV i.e. -3.4 eV, and so on.

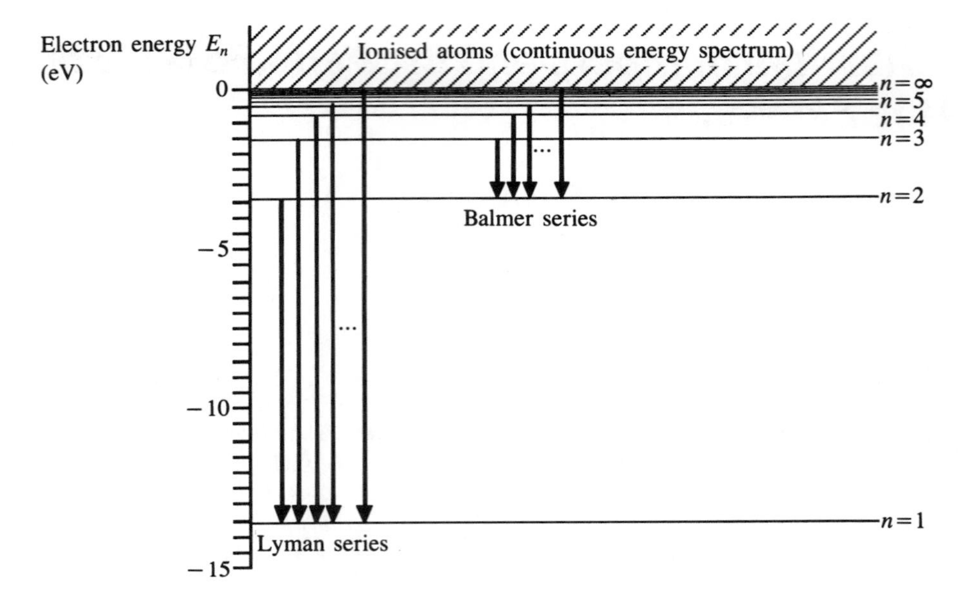

Figure 2.1 *Energy-level diagram for the hydrogen atom.*

Figure 2.1 shows that only the energy levels below $E_n = 0$ are quantized. The energy of any electrons in the atom (normally one, in the case of hydrogen) must coincide with the horizontal lines (or energy levels) shown in the figure. Even though the energy levels with n larger than about 10 are very closely spaced, the levels are still very discrete. An electron cannot have energies which lie between these levels. The regions between levels are called forbidden gaps, or energy gaps, for this reason. It's worth pointing out before you go any further that these so-called gaps are *energy* gaps — they're *not* gaps in space. Students sometimes forget this and it causes misunderstanding later on.

If an electron is torn free from the atom and is at rest (i.e. not moving) its energy is zero: this is the meaning of the $n = \infty$ level. It's possible for this so-called 'free' electron to have energy greater than zero if it's mobile, in which case it would be in the continuum of levels shown with positive energies (i.e. values of E_n above zero in the figure). In this region the energy of the electrons is not quantized and the electron can have any value of kinetic energy. Note, by the way, that Bohr's model doesn't account for these positive values of E_n, but it does show in equation (2.1) that $E_n = 0$ for $n = \infty$ because dividing by ∞ always gives a result of 0.

Figure 2.1 can be used to illustrate the origin of the various spectral series observed in the emission spectrum of hydrogen. According to Bohr, the atom emits light when one of its electrons falls from one of the higher energy levels to a lower level. In his model, when the electron is in the $n = 4$ energy level, the electron is in the fourth circular orbit. If the electron falls to the 2nd orbit ($n = 2$), for example, the atom loses energy $E_4 - E_2$, and this energy is radiated as a quantum of light (i.e. a photon), with frequency given by

$$hf_{42} = E_4 - E_2 \tag{2.2}$$

This energy difference is shown in Fig. 2.1; it's represented by the arrow extending from the $n = 4$ energy level to the $n = 2$ energy level. For the one electron in hydrogen the lowest level ($n = 1$) is the ground state.

2.1.1 Shortcomings of Bohr's model

Bohr's model was rather poor in some respects. A major disadvantage was that it didn't accurately describe large, more complex atoms. It was good enough for hydrogen, but hydrogen is the smallest atom and is therefore relatively simple in structure. Bohr worked with Arnold Sommerfeld (1868–1951), another great scientist, to improve his basic model. They suggested that there should be more quantum numbers. Remember that Bohr had already devised the principal quantum number n in his third postulate (equation (1.8)):

$$m_o v r_n = \frac{nh}{2\pi}$$

Bohr and Sommerfeld suggested that the principal quantum number n should have two more quantum numbers associated with it: l and m, making a set of three quantum numbers to describe an electron, such that:

1. The principal quantum number n denotes the energy of the electron.
2. The orbital quantum number l denotes the angular momentum of the electron.
3. The magnetic quantum number m denotes the way the electron behaves in a magnetic field.

These extra quantum numbers changed the simple Bohr model of the atom in a fundamental way. In the Bohr model the electron moves in a circular orbit (denoted by n) around the nucleus. The addition of the two new quantum numbers meant that the orbit denoted by n could be split into more orbits, all denoted by the same value of n, and that they need not be circular — they can be elliptical; the shapes of all these orbits are determined by their l values.

RESEARCH 2.1

Look up the definition of an ellipse, and write that definition in your notebook.

Hence the atomic nucleus becomes the centre of a circular orbit and the focus of an elliptical orbit (Figure 2.2). For a particular value of n, then, the orbits will all have the same energy E_n but different shapes denoted by their l values. (We'll consider the numerical values these different quantum numbers can have later.) Different orbits for the same total energy E_n are said to be *degenerate* orbits.

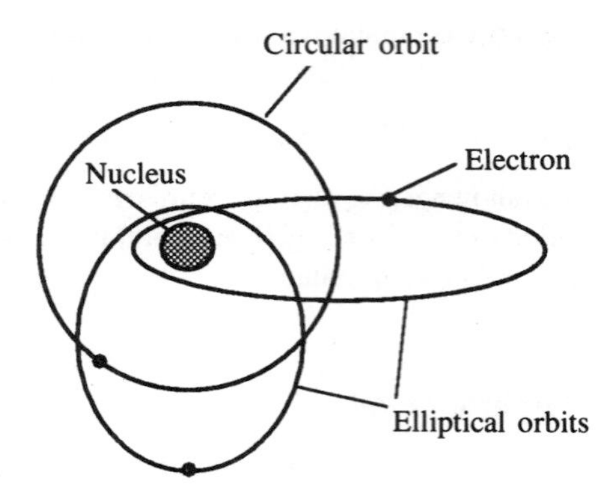

Figure 2.2 *Representation of an atom with circular and elliptical orbits.*

Later on, in 1925, Uhlenbeck and Goudsmit proposed a fourth quantum number called the spin quantum number, s. The need for the spin quantum number arises if you think of an electron like a tiny billiard ball spinning on an axis: the ball can spin in either a clockwise or an anticlockwise direction. Hence an electron can have one of two different spins. These spin directions are shown in various ways in the textbooks, for example + and −, or up and down, or ↓ and ↑, or left and right. In this book we'll use $s = +\frac{1}{2}$ or $s = -\frac{1}{2}$.

To summarize, there are four quantum numbers:

1 principal quantum number n;
2 orbital quantum number l;
3 magnetic quantum number m;
4 spin quantum number s.

No two electrons in an atom can have the same set of four quantum numbers, i.e. a set of these four quantum numbers describes an *individual* electron in a *single* atom. This is called the Pauli exclusion principle after an explanation given by Wolfgang Pauli (1900–1958) in 1924. Pauli was awarded the Nobel Prize for Physics in 1945.

Pauli's exclusion principle can be stated as follows:

> In an atom containing more than one electron, no two electrons can be in the same state, characterized by the same four quantum numbers n, l, m and s.

If you've looked in some of the text books you may not recognize the symbols n, l, m and s for the four quantum numbers; don't worry about this — authors are notoriously inconsistent in their use of notation and you may see, for example, n, n_l, n_m and n_s representing them. *It doesn't really matter what you use so long as you define the meanings of the symbols and so long as you're consistent.*

SELF-ASSESSMENT QUESTION 2.1

Cover up the previous few paragraphs and list the names of the quantum numbers in your notebook, giving a suitable symbol for each.

What do I mean by 'a set of four quantum numbers'? Well, I've already mentioned that quantum numbers have numerical values, and it's these four numerical values, one corresponding to each quantum number, that make up a set. How do we know what values to choose to make up a set? Well, this is made easy for us because there are mathematical rules which we can apply.

Before I show you what these rules are and how to apply them, it's a good idea to remind you what we're doing. We've seen how an electron in an atom has to occupy a particular orbit or energy level such that the electron has energy E_n. The number of this orbit is n, the principal quantum number. Bohr's model really only describes hydrogen and so more quantum numbers were introduced to produce a model which describes larger atoms which contain more electrons. A single atom of any substance contains one or more electrons. Each of these electrons can be described or identified by a unique set of four quantum numbers, such that no other electron *in that atom* can have the same set.

2.2 The Periodic Table of the Elements

Before we can pursue the ideas of quantum theory we need to consider the Periodic Table of the Elements. A version of the Table is shown in Table 2.1. The figure in the top left-hand corner of each box is the atomic number (symbol Z); this is equal to the number of electrons in a neutral atom of the element indicated.

SELF-ASSESSMENT QUESTION 2.2

How many electrons are there in neutral atoms of the following elements?

carbon C	oxygen O	nitrogen N
silicon Si	germanium Ge	iron Fe
neon Ne	chlorine Cl	mercury Hg
silver Ag	potassium K	yttrium Y
lead Pb	tungsten W	

The Periodic Table is a tabular arrangement of all the known elements in order of increasing atomic number Z. It is important because it gives a tremendous amount of information about the properties of all the known elements. All technological developments are materials-based, so a knowledge of the properties of those materials or, more importantly, the ability to determine their properties from available information, is extremely important to scientists and engineers. The rows across the table, of which there are seven, are known as periods and the columns as groups. There are eighteen groups from **IA** to **0** (or *1* to *18*). The first method of labelling was introduced by the International Union of Pure and Applied Chemistry (IUPAC) earlier this century but was replaced by the simpler method of sequential numbering from 1 to 18 when the IUPAC met in 1992.

The elements in a group have similar chemical behaviour because of a similarity in their electronic configurations. The bold labels showing roman numerals are the

Table 2.1 ***The Periodic Table of the Elements, showing the pre-1992 IUPAC group names (IA–0) and the post-1992 IUPAC group names (1–18)***

IA	IIA	IIIA	IVA	VA	VIA	VIIA	←	VIII	→	IB	IIB	IIIB	IVB	VB	VIB	VIIB	0
1	*2*	*3*	*4*	*5*	*6*	*7*	*8*	*9*	*10*	*11*	*12*	*13*	*14*	*15*	*16*	*17*	*18*
1 H																	2 He
3 Li	4 Be											5 B	6 C	7 N	8 O	9 F	10 Ne
11 Na	12 Mg											13 Al	14 Si	15 P	16 S	17 Cl	18 Ar
19 K	20 Ca	21 Sc	22 Ti	23 V	24 Cr	25 Mn	26 Fe	27 Co	28 Ni	29 Cu	30 Zn	31 Ga	32 Ge	33 As	34 Se	35 Br	36 Kr
37 Rb	38 Sr	39 Y	40 Zr	41 Nb	42 Mo	43 Tc	44 Ru	45 Rh	46 Pd	47 Ag	48 Cd	49 In	50 Sn	51 Sb	52 Te	53 I	54 Xe
55 Cs	56 Ba	57 La	72 Hf	73 Ta	74 W	75 Re	76 Os	77 Ir	78 Pt	79 Au	80 Hg	81 Tl	82 Pb	83 Bi	84 Po	85 At	86 Rn
87 Fr	88 Ra	89 Ac	104 Rf	105 Ha	106 Sg	107 Ns	108 Hs	109 Mt	110								

58 Ce	59 Pr	60 Nd	61 Pm	62 Sm	63 Eu	64 Gd	65 Tb	66 Dy	67 Ho	68 Er	69 Tm	70 Yb	71 Lu	**IIIA**
90 Th	91 Pa	92 U	93 Np	94 Pu	95 Am	96 Cm	97 Bk	98 Cf	99 Es	100 Fm	101 Md	102 No	103 Lr	*3*

Source: Lide, D.R. (Ed.), *CRC Handbook of Chemistry and Physics*, 75th edn., CRC Press, 1994

groups, e.g. carbon (C) is in group IVB, silicon (Si) is in group IVB, sodium (Na) is in group IA, and so on.

RESEARCH 2.2
Find another Periodic Table and compare it with the one shown in Table 2.1. You may find that it is slightly different to the one here because authors tend to be inconsistent about where they put hydrogen (H), and whether or not they put the noble gases (helium He to radon Rn) in group 0. Note any differences in your notebook; also note the reference so that you can find the same table again. Also note if the group numbers conform to the new IUPAC system.

So, we've established that the Periodic Table is useful, although you may be a bit sceptical about its use at this stage. We can put it to use immediately, however, by considering the electronic configurations of some of the elements shown in it. By electronic configuration, I mean a description of all the electrons in the atom according to our atomic model which uses four quantum numbers. Now is the time to tell you what the simple mathematical rules are that I referred to earlier. These rules help us to answer the question 'How do the electrons arrange themselves in an atom?'

2.3 How to fill an atom with electrons

There are eight rules altogether:

1. A neutral atom has a number of electrons equal to its atomic number Z (this is where you refer to the Periodic Table).
2. In an unexcited (or neutral) atom, the electrons are in their lowest possible (i.e. ground) states.
3. No two electrons in an atom can have the same four quantum numbers (i.e. Pauli's exclusion principle applies).
4. n may take positive, non-zero integer values, i.e. $0 < n$.
5. The maximum number of electrons in an atom that can have the value n is $2n^2$.
6. l may be zero or a positive integer, such that $l \leq n - 1$.
7. $m \leq \pm l$, so m can be negative, zero or positive.
8. $s = \pm \frac{1}{2}$, so there are only two choices for s.

The spin quantum number s determines the magnetic nature of atoms. Hund deduced from measurements of the magnetic moment of atoms that levels are filled first by electrons having the same value of s, $+\frac{1}{2}$, and when the sub-shell denoted by l is half-full, the remaining states would be filled by electrons with $s = -\frac{1}{2}$. This means that the electrons are arranged to give the maximum possible total spin in an individual atom. This means that, to get the maximum value of s, the first electrons going into a level denoted by a particular value of l must have spin $= +\frac{1}{2}$. This is called Hund's rule.

Let's consider hydrogen.

The Periodic Table tells us that hydrogen H has an atomic number of 1, therefore according to rule (1) a neutral atom of hydrogen has one electron (of course, we know this already). Now we need to determine the electronic configuration of that atom, so we tabulate the values that n, l, m and s can take for that one electron.

Rule (2) tells us that this one electron in hydrogen will be in a level denoted by $n=1$ (the lowest possible value of n) and corresponding lowest values of l, m and s, i.e. the electron will be in its lowest possible energy state. (We know that this energy is equal to -13.6 eV.)

Rule (3) can't be violated for hydrogen because there's only one electron.

Rule (4) tells us that n is a positive non-zero integer. Always choose the lowest available value for n, so it's 1 for hydrogen.

For $n=1$, $2n^2=2$ (rule (5)). So we can have a maximum of two electrons for which $n = 1$. This isn't a problem for hydrogen because there's only one electron anyway.

Rule (6) tells us that l is less than or equal to $n-1$. For hydrogen n is 1, so $n-1$ is 0. l can't be lower than this, so $l=0$. Note that l cannot equal 1 here because of rule (2)— always use the lowest available value.

According to rule (7), $m = 0$. Note that m can take negative values of l, but in this case l is zero anyway, so m cannot be negative.

From rule (8) and Hund's rule, $s = +\frac{1}{2}$.

So now we can write the electronic configuration of hydrogen as shown in Table 2.2.

Table 2.2 *The electronic configuration of hydrogen*

Element	*Electronic configuration*				
	Z	*n*	*l*	*m*	*s*
H	1	1	0	0	$+\frac{1}{2}$

Now let's consider the next element in the Periodic Table, helium (He). Helium has an atomic number of 2, so we need to put in two electrons. The first electron that goes into the helium atom has the same configuration as that for hydrogen. What's the configuration of the second? Apply the rules as previously:

- n is still 1 because it's the lowest non-zero positive value the electron can take and it doesn't contravene the $2n^2$ rule (rule 5);
- l must therefore be 0;
- m must therefore be 0;
- s must be $-\frac{1}{2}$ because the shell $n = 1$ is already half full; also, Pauli's exclusion principle must be maintained.

The electronic configuration of helium is shown in Table 2.3.

Table 2.3 *The electronic configuration of helium*

Element		*Electronic configuration*			
	Z	*n*	*l*	*m*	*s*
He	2	1	0	0	$+\frac{1}{2}$
		1	0	0	$-\frac{1}{2}$

The next element in the Periodic Table is lithium Li, which has three electrons. Two of these electrons will have the same configurations as those of the electrons in helium. We therefore need only determine the configuration of the third electron. Rule (5) tells us that only two electrons can go in levels for which $n = 1$ (because $2n^2 = 2$ when $n = 1$), so the third electron must go into a level with a different value of n. Application of rules (2) and (4) means that the new principal quantum number for the third electron must be $n = 2$. If $n = 2$, l can have values 0 or 1 because $l \leq n - 1$. In fact $l = 0$ because of rule (2). Rule (7), $m \leq \pm l$, means that $m = 0$. Lastly, $s = +\frac{1}{2}$. The electronic configuration of lithium is shown in Table 2.4.

Table 2.4 *The electronic configuration of lithium*

Element		*Electronic configuration*			
	Z	*n*	*l*	*m*	*s*
Li	3	1	0	0	$+\frac{1}{2}$
		1	0	0	$-\frac{1}{2}$
		2	0	0	$+\frac{1}{2}$

SELF-ASSESSMENT QUESTION 2.3

Try the next element, beryllium Be, for yourself. Draw a table listing *Z, n, l, m* and *s* to help you.

SELF-ASSESSMENT QUESTION 2.4

Try another for yourself. Tabulate the configurations of all the electrons in the next element, boron B.

We could go on putting in electrons in this manner to make larger and larger atoms. In fact, there's a hitch after argon Ar ($Z = 18$) because in larger atoms the inner electrons screen the outer electrons from the nucleus, resulting in a more complex filling sequence where the outer energy levels may overlap, causing the levels to be filled out of the logical order. So for atoms for which $Z > 18$, the eight rules listed above no longer all apply. The transition elements in the middle block of the Periodic Table are particularly affected by this screening effect.

2.4 The shorthand method of writing electronic configuration

As you may imagine, writing out electronic configurations by tabulating all the quantum-number values for every electron is very laborious. This tedium is avoided by using what I call the 'shorthand' method; if you've studied chemistry you've probably come across it before. This notation depends on the value of l (Table 2.5).

Table 2.5 *Notation used for the orbital quantum number l*

Value of l	*Designatory letter*
0	s
1	p
2	d
3	f

SELF-ASSESSMENT QUESTION 2.5

What is the maximum number of s electrons, p electrons, d electrons and f electrons that may exist in a given shell?[1]

All we need to know to use this notation is the principal quantum number n and how many electrons have a particular value of s, p, d or f. For example, the electronic configuration of hydrogen is $1s^1$ according to this notation because it has only one electron described by $n = 1$ and $l = 0$. (Refer back to Table 2.2 to find out what the full, tabulated configuration is.) The first 1 denotes the principal quantum number n and the superscript 1 shows how many electrons have $l = 0$ in the shell denoted $n = 1$. So what's the shorthand notation for helium? It's $1s^2$, because helium has two electrons, both designated by $n = 1$ and $l = 0$. We say that helium has two 1s electrons. The shorthand notation for lithium is $1s^2 2s^1$. Do you see why? (Refer back to the tabulated version in Table 2.4.) Table 2.6 is a list of the shorthand configurations of neutral atoms in the ground state for all the elements from $Z = 1$ to $Z = 18$. Notice that the sum of the superscript values for each element is equal to the atomic number Z for that element. Also, there is no information about the spin quantum number s. This can be deduced, if necessary, by applying Hund's rule and Pauli's exclusion principle.

SELF-ASSESSMENT QUESTION 2.6

Write the shorthand electronic configurations of the following ions:

$$C^{4+} \ O^{2-} \ Na^{+} \ Cl^{-}$$

Iron's electronic configuration

What is it about iron that makes it so special? Why is it magnetic? If we look at the electronic configuration of iron we'll find out. Look at the Periodic Table (Section 2.2) to find out how many electrons there are in an atom of iron (Fe).

1 Note that a shell is denoted by a particular value of n, such that any electron with $n=1$ is in the first shell, any electron with $n=2$ is in the second shell, and so on. This means that the $2n^2$ rule determines the maximum number of electrons allowed in a shell.

Table 2.6 *Shorthand electronic configurations of the first 18 elements*

Z	*Element*		*Electronic configuration*
1	H	hydrogen	$1s^1$
2	He	helium	$1s^2$
3	Li	lithium	$1s^22s^1$
4	Be	beryllium	$1s^22s^2$
5	B	boron	$1s^22s^22p^1$
6	C	carbon	$1s^22s^22p^2$
7	N	nitrogen	$1s^22s^22p^3$
8	O	oxygen	$1s^22s^22p^4$
9	F	fluorine	$1s^22s^22p^5$
10	Ne	neon	$1s^22s^22p^6$
11	Na	sodium	$1s^22s^22p^63s^1$
12	Mg	magnesium	$1s^22s^22p^63s^2$
13	Al	aluminium	$1s^22s^22p^63s^23p^1$
14	Si	silicon	$1s^22s^22p^63s^23p^2$
15	P	phosphorus	$1s^22s^22p^63s^23p^3$
16	S	sulphur	$1s^22s^22p^63s^23p^4$
17	Cl	chlorine	$1s^22s^22p^63s^23p^5$
18	Ar	argon	$1s^22s^22p^63s^23p^6$

SELF-ASSESSMENT QUESTION 2.7

So, how many electrons are there in iron?

Twenty-six electrons therefore have to be put into an atom to make iron. The electronic configuration is interesting. Iron is a transition element, meaning that its outer shells overlap because of the large size of the atom. This means the electrons won't conform to the rules we saw earlier (Section 2.3). The electronic configuration is

$$1s^22s^22p^63s^23p^63d^64s^2$$

Both shells for $n = 1$ and $n = 2$ are full (remember the $2n^2$ rule!).

SELF-ASSESSMENT QUESTION 2.8

What is the maximum number of electrons that will go in the shell $n = 3$?

Look at the shorthand electronic configuration of iron again.

The third shell contains 14 electrons, then 2 electrons go into the fourth shell. Consequently the third shell is short of 4 electrons. To be specific, the third shell is short of 4 d electrons (there can be a maximum of 10 d electrons in a shell).

SELF-ASSESSMENT QUESTION 2.9

Why is there a maximum of 10 d electrons in a shell?

Consider the electronic configuration of iron in full (Table 2.7). The 3d electrons are unusual because four of them (in bold type) are not paired, i.e. they have no spin partner. It's this resulting spin value that is important. Normally two electrons are paired in a state, such that one of the pair has spin $-\frac{1}{2}$ and the other has spin $+\frac{1}{2}$. When added together, $-\frac{1}{2} + +\frac{1}{2}$ is zero. Hence for a so-called *spin pair* of electrons, the total spin is zero. In iron, four 3d electrons are unpaired, giving a total spin of $4 \times +\frac{1}{2} = +2$ (Table 2.8). It's this large total spin value which is responsible for iron's magnetic properties. Each unpaired electron produces a magnetic moment of one Bohr magneton (μ_B), so iron has a magnetic moment of 4 Bohr magnetons. Each Bohr magneton contributes 9.27×10^{-24} A m^2 to the atom's magnetic moment, given by

$$\mu_B = \frac{he}{4\pi m_o}$$
$$= 9.27 \times 10^{-24} \text{A m}^2$$

The Bohr magneton is the smallest theoretically possible magnetic moment.

Table 2.7 *The electronic configuration of iron*

	n	*l*	*m*	*s*	
1	1	0	0	$+\frac{1}{2}$	1s electrons
2	1	0	0	$-\frac{1}{2}$	
3	2	0	0	$+\frac{1}{2}$	2s electrons
4	2	0	0	$-\frac{1}{2}$	
5	2	1	−1	$+\frac{1}{2}$	2p electrons
6	2	1	0	$+\frac{1}{2}$	
7	2	1	+1	$+\frac{1}{2}$	
8	2	1	−1	$-\frac{1}{2}$	
9	2	1	0	$-\frac{1}{2}$	
10	2	1	+1	$-\frac{1}{2}$	
11	3	0	0	$+\frac{1}{2}$	3s electrons
12	3	0	0	$-\frac{1}{2}$	
13	3	1	−1	$+\frac{1}{2}$	3p electrons
14	3	1	0	$+\frac{1}{2}$	
15	3	1	+1	$+\frac{1}{2}$	
16	3	1	−1	$-\frac{1}{2}$	
17	3	1	0	$-\frac{1}{2}$	
18	3	1	+1	$-\frac{1}{2}$	
19	3	2	−2	$+\frac{1}{2}$	3d electrons
20	**3**	**2**	**−1**	$+\frac{1}{2}$	
21	**3**	**2**	**0**	$+\frac{1}{2}$	
22	**3**	**2**	**+1**	$+\frac{1}{2}$	
23	**3**	**2**	**+2**	$+\frac{1}{2}$	
24	3	2	−2	$-\frac{1}{2}$	
25	4	0	0	$+\frac{1}{2}$	4s electrons
26	4	0	0	$-\frac{1}{2}$	

Table 2.8 *Spin states of iron*

Principal quantum number *n*						
n=1	*n*=2		*n*=3			*n*=4
1s	2s	2p	3s	3p	3d	4s
↑	↑	↑↑↑	↑	↑↑↑	↑↑↑↑↑	↑
↓	↓	↓↓↓	↓	↓↓↓	↓	↓

2.5 **Properties of the elements**

The physical and chemical properties of the elements are determined by their positions in the Periodic Table which, in turn, are determined by their electronic configurations. Table A1 (Appendix A) lists the electronic configurations of the neutral (i.e. non-ionized) atoms of all the known elements from $Z = 1$ to $Z = 107$.

At the time of completing this book the international chemistry community is in uproar about the element for which $Z = 106$, previously called unnilhexium and given the symbol Unh. The discovery of unnilhexium was reported in 1974, but is still waiting to be officially named by the International Union of Pure and Applied Chemistry (IUPAC). The American Chemical Society (ACS) has its own Committee on Nomenclature and has recommended the adoption of the name seaborgium[2] (symbol Sg) for unnilhexium. There is enormous controversy over this issue: the IUPAC had proposed the name rutherfordium (symbol Rf) for this element, but this has not been a popular choice.

Other large elements have been discovered and reported in recent years (from $Z = 107$ in 1976 to $Z = 110$ in 1994). Like rutherfordium ($Z = 104$) and hahnium ($Z = 105$), which were temporarily called unnilquadium (Unq) and unnilpentium (Unp) respectively, they have to await an official name which is approved by IUPAC. The recommended names adopted by the IUPAC and the ACS are shown in Table 2.9.

Table 2.9 ***The controversial names for some of the new elements, proposed by the IUPAC and the ACS***

Z	*IUPAC name and symbol*		*ACS name and symbol*	
104	dubnium	Db	rutherfordium	Rf
105	joliotium	Jl	hahnium	Ha
106	rutherfordium	Rf	seaborgium	Sg
107	bohrium	Bh	nielsbohrium	Ns
108	hahnium	Hn	hassium	Hs
109	meitnerium	Mt	meitnerium	Mt

2 Named after the Nobel Laureate Glenn T. Seaborg, who co-discovered nine elements including plutonium.

Not many of these large atoms have been detected because they are very unstable and have to be artificially created. These very large atoms are all radioactive (from $Z = 89$), as are technetium Tc (43), promethium Pm (61), francium Fr (87) and radium Ra (88).

Physical and chemical properties are determined by the valency of the element. Valency is equivalent to the number of electrons in the outer sub-shell, or to the number of available states in the outer sub-shell. Valency is therefore related to the ease with which an atom may become ionized by gaining or losing outer electrons. For example, consider oxygen. It has 6 electrons in its outer shell denoted by $n = 2$, leaving space for two more. It is therefore in group VI of the Periodic Table and is divalent (it has a valency of two) and will combine with two other electrons from at least one other atom to form compounds.

Each period (horizontal row) in the Periodic Table exhibits increasing electronegativity to the right. The most electropositive elements (i.e. those which form positive ions) are in group I whereas the most electronegative elements (i.e. those which form negative ions) are in group VIIB. Hence the most reactive elements can be located in groups IA and VIIB of the Periodic Table. The elements in group IA from lithium to francium all have one s electron in their outer shell, making it very easy for these elements to lose that one electron and become a positive ion. These elements are called the alkali metals because they have alkaline properties and are good conductors. The elements of group VIIB have seven electrons in their outermost shell (two s electrons and five p electrons), making it very easy to acquire one more electron to become a negative ion. These elements are the halogens. They react very easily with the alkali metals because, by sharing their outermost electrons, both ions have a complete set of eight electrons in their outer shells (two s, six p).

The least reactive elements are those in group 0: the noble gases, helium He to radon Rn. These all have two s electrons and six p electrons in their outer shells except helium, of course, which has the two s electrons only, so are very stable.

The elements in the middle block of the Table from group IIIA to group IIB are the transition metals. This is a large group of elements, including most of the commonly used metals. All these elements show considerable similarities with their horizontal neighbours in the Periodic Table. They have several properties in common including:

- their appearance (hard, brittle, reflecting surface when polished);
- the ability to form alloys;
- the ability to be hammered into shape (malleability);
- the ability to be drawn into wire (ductility);
- high melting point;
- good thermal conductivity;
- multiple valencies (e.g. iron has valencies 2 or 3); and
- the ability to form coloured compounds.

The transition metals are so-called because they are large enough for their outer subshells to overlap. This causes some interesting properties, ferromagnetism for example. The three adjacent elements in the fourth period of group VIII, iron, cobalt and nickel, are all magnetic.

2.6 Summary

Bohr's model predicted quantized energy levels denoted by the principal quantum number n in the hydrogen atom but wasn't sophisticated enough to describe what happens in larger atoms, so a set of four quantum numbers was devised to enable every electron in an atom to have its own unique configuration denoted by the four quantum numbers n, l, m and s. This meant that a single shell in the atom could contain more than one sub-shell or orbit, and these orbits could be elliptical or circular. A set of mathematical rules was devised so that these modifications to Bohr's basic model gave accurate descriptions of electronic behaviour in larger atoms. The electronic structure of the elements in the Periodic Table was then understood, and it was found that the chemical behaviour of these elements is largely determined by their electronic configurations and that elements in the same group have similar properties. The first eighteen elements, hydrogen to argon, are described exactly by simple application of the rules, whereas larger atoms show some anomalies because of their large size. We've seen that the elements can be described by tabulating the four quantum numbers for each electron in an atom of the element, and that the labour involved in such an exercise can be reduced by using a shorthand method based on a set of symbols (s, p, d, f) representing l, the orbital quantum number.

The four quantum numbers lead to the electronic configurations of the elements and explain the structure of the Periodic Table. They are therefore fundamentally important when explaining and predicting the chemical and physical properties of all the elements.

2.7 Tutorial question

2.1 Without referring to this book, tabulate the electronic configurations of all the electrons in argon (Ar), showing the values of all four quantum numbers.

2.8 Suggested further reading

Anderson, J. C., Leaver, K. D., Rawlings, R. D. and Alexander, J. M., *Materials Science*, 3rd edn, Van Nostrand Reinhold, 1985, pp. 61–65.

Jerrard, H. G. and McNeill, D. B., *Dictionary of Scientific Units*, 6th edn, Chapman and Hall, 1992, p. 199.

Lide, D. R. (Ed.), *CRC Handbook of Chemistry and Physics*, 75th edn, CRC Press, 1994.

Rouvray, D., Elementary, my dear Mendeleyev, *New Scientist*, **141** 36–39, 1994.

Seaborgium lacks 'perspective of history'. *Chemistry and Industry*, 17 October 1994, p. 799.

Solymar, L. and Walsh, D., *Lectures on the Electrical Properties of Materials*, 3rd edn, Oxford University Press, 1986, section 4.4 The periodic table, pp. 72–77.

The first superheavy element weighs in, *Chemistry and Industry*, 5 December 1994, p. 929.

CHAPTER 3

Solid materials

Aims and objectives

Most modern electronic devices are made of solids, so a model describing what happens to electrons in solids is necessary to explain the electrical behaviour of the devices. The energy-band model is introduced in this chapter and used to classify conductors, insulators and semiconductors. Electrical conduction in these solids is briefly described using the band model. The common semiconductors are crystalline; the crystal structure of these will be described, and a method used for defining crystal planes will be explained. The chapter finishes by listing some elemental and compound semiconductors and classifying them according to their Periodic-Table groups.

3.1 Energy bands in solids

So far we've been cotncerned with individual atoms and their well-defined energy levels or shells. This model is fine for looking at atoms which are far apart, such as a gas. What happens when we bring these atoms closer together, as in a solid?

3.1.1 Formation of energy bands

When atoms are far apart, as in a gas (Figure 3.1), the energy levels of any given atom are identical to those described in Chapters 1 and 2. As the internuclear spacing is reduced (i.e. as the atoms are brought closer together) a stage is reached where the outer valence electrons begin to interact with each other, and the gas liquefies (Figure 3.2). The external fields produced by the electrons of neighbouring atoms cause the valence levels to split into finely spaced levels, each of which has a slightly different energy value. As the atoms are compressed further the liquid forms a solid (Figure 3.3) and the interatomic spacing has been further reduced so that the inner energy levels also start to split.

The result of bringing atoms closer together is that discrete energy levels occurring in the gas split into a range of sublevels in the solid that are so close together that they form a band of energy states (Figure 3.4), called an energy band. These energy states are so close that they can be considered continuous.

Figure 3.5 shows how a convenient energy-band diagram can be drawn to represent

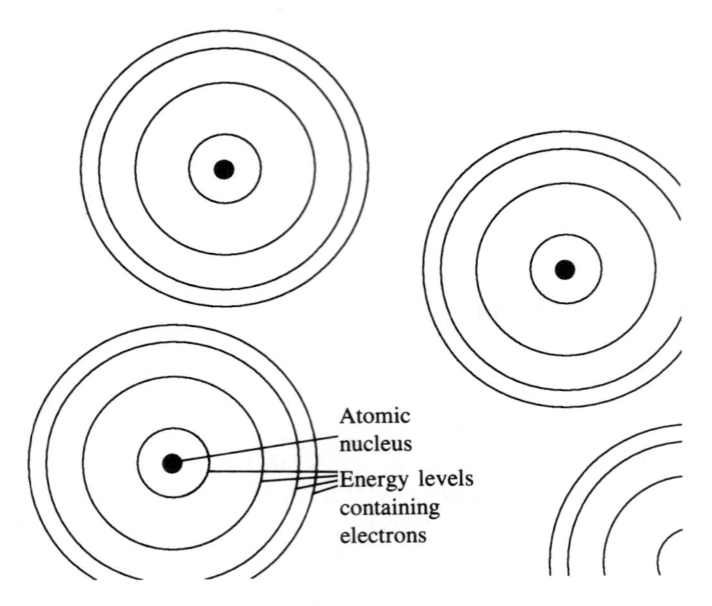

Figure 3.1 *Gaseous atoms.*

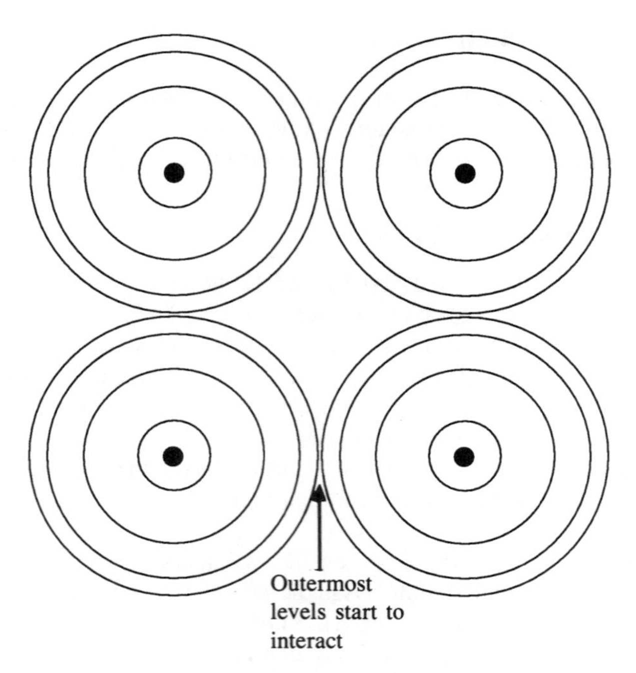

Figure 3.2 *Four atoms brought close together.*

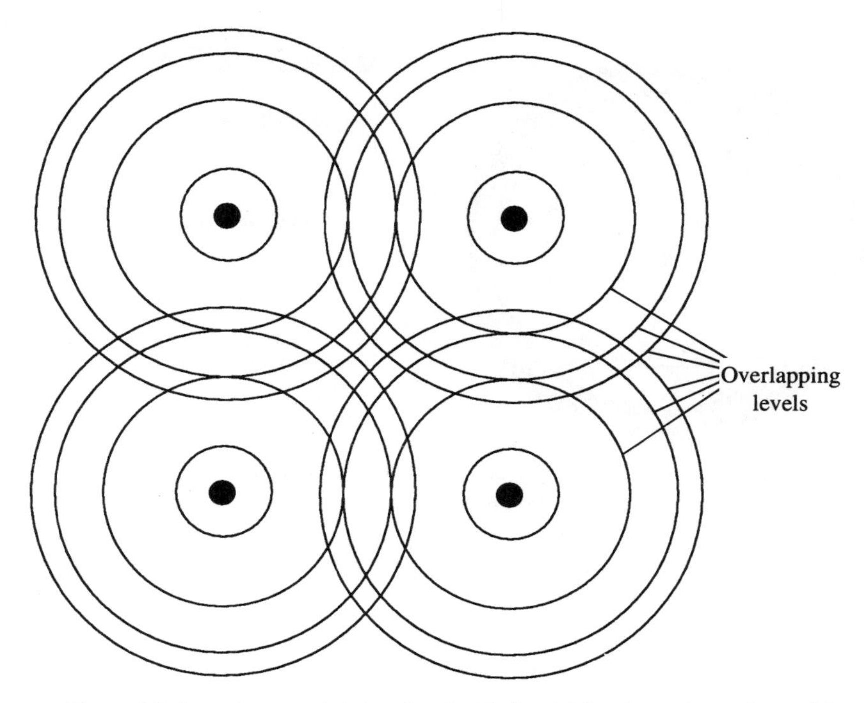

Figure 3.3 *Four atoms so close together that their outer levels overlap, as in a solid.*

the energy states in one of the atoms of the solid. All the atoms making up the solid are represented by the same diagram. Electron energy is plotted vertically such that the arrow indicates increasingly positive energy (or, if you prefer, decreasingly negative energy). The horizontal axis is just the distance through the material, but this is never

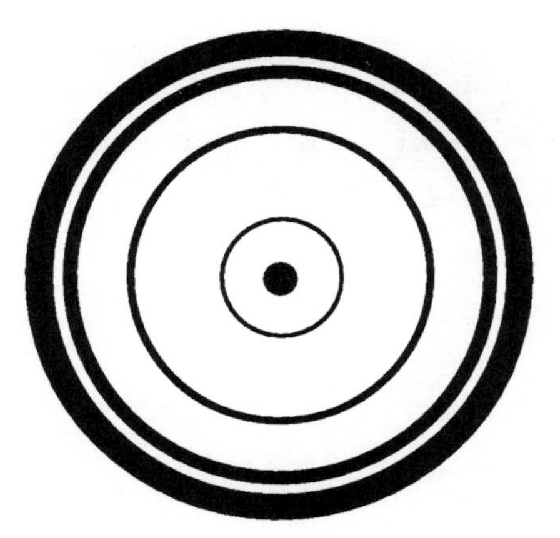

Figure 3.4 *Schematic representation of one of the four atoms in the solid, showing broadening of the outer levels.*

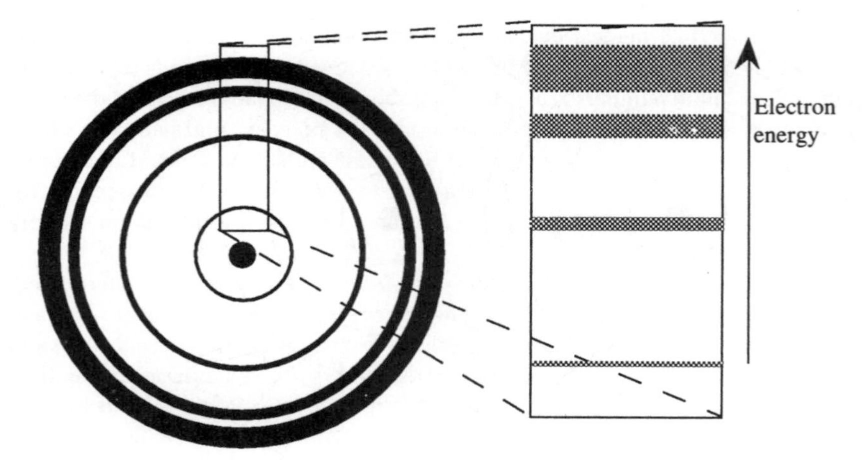

Figure 3.5 *A section through one of the atoms in the solid, showing how the broadened bands can be shown in the form of an energy-band diagram.*

specified. Movement along the horizontal axis in an energy-band diagram indicates that electrons are moving from atom to atom.

Sodium — a typical conductor

Consider sodium as an example. Sodium is a good conductor of electricity and is therefore quite an interesting element to study. A conductor is a material that allows its electrons to move freely through it. Figure 3.6 shows the energy levels of sodium

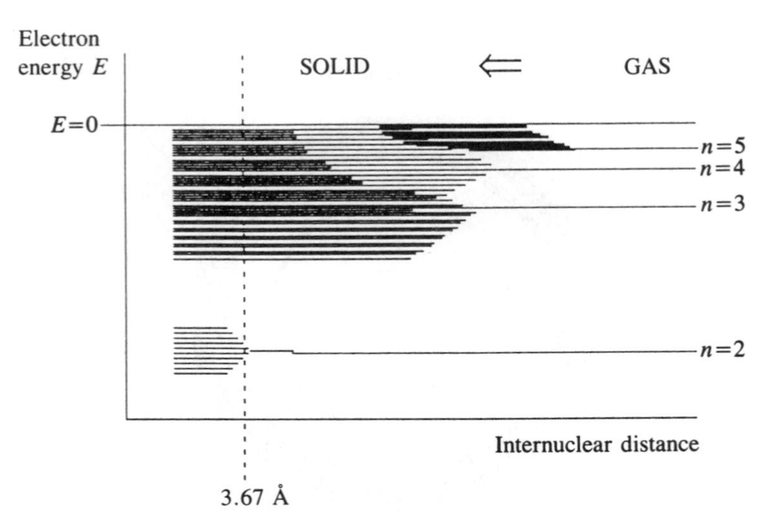

Figure 3.6 *A sketch showing the broadening of discrete energy levels into bands as the interatomic spacing in sodium is decreased.*

plotted against the internuclear distance of the sodium atoms. Notice how the levels broaden into bands as the atoms are brought closer together. The bands representing the principal quantum numbers 2, 3, 4 and 5 are shown. The band for $n = 1$ has been left out for convenience. The equilibrium separation of the atoms in solid sodium is 3.67 Å as indicated in the figure. This means that the average distance apart of the atomic nuclei in solid sodium is 3.67 Å when the atoms are in a stable state. We can see that, for an internuclear distance of 3.67 Å, the $n = 3$ and $n = 4$ bands in solid sodium actually overlap.

Figure 3.7 shows the energy-band diagram for solid sodium at the internuclear distance of 3.67 Å. This shows clearly how the energy levels corresponding to $n = 3$ and $n = 4$ overlap at this internuclear distance. These two bands overlap to such an extent that both bands extend into the *continuum*, where the energy values are positive. If an electron reaches the continuum it means it has left the material altogether, and is free to roam about in the atmosphere above the material with a positive value of kinetic energy. Electrons emitted from the material in the photoelectric experiment are examples of this (see Section 1.5.1).

The separation between the discrete levels is so small that the band of sublevels can be considered to be continuous. As with isolated atoms, the Pauli exclusion principle still applies. One electron can occupy each discrete sublevel in an energy band. The lowest-energy sublevel is occupied first, then the next highest, and so on, until all the electrons are accommodated.

Remember that an energy band is nothing more than a collection of energy levels which may or may not contain electrons. The interesting property of a band, however, is the closeness of these energy levels in the band — they're so close that we could consider the range of energies within the band to be continuous. Be careful here —

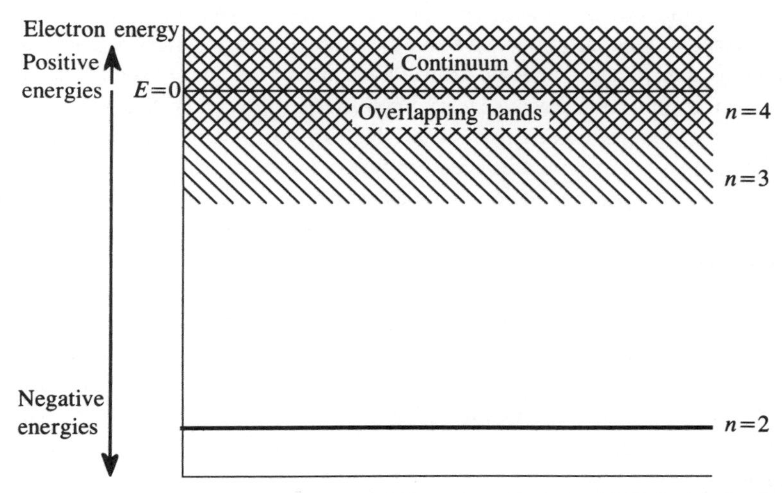

Figure 3.7 *Energy bands for solid sodium at an internuclear distance of 3.67 Å, showing how the outer bands overlap and extend into the continuum.*

don't confuse the energies in the band with the *continuum* of positive energies above $E = 0$ which represent electrons which have left the material completely.

Let's consider sodium in a bit more detail. Why is it a good conductor? It is an alkali metal from group I of the Periodic Table with an atomic number $Z = 11$, and has the electronic configuration $1s^2 2s^2 2p^6 3s^1$, with the important single 3s electron which is responsible for its good conductivity. Two electrons *per atom* are in the band denoted by $n = 1$. Eight electrons are in the band denoted by $n = 2$. Only one electron per atom has a ground state in the third band.

Consider this one electron per atom in the $n = 3$ band in sodium. In order to conduct electricity (i.e. form a current), this electron has got to be able to wander through the material. The lone electron in the third band will be at the bottom of the band at absolute zero because that is the lowest energy state it can occupy. Above it there are empty states for which $n = 3$ and the empty states for which $n = 4$. The electron cannot move to a nearby atom (i.e. conduct) because the state it would like to occupy already has an electron in it. So how can the electron move up into empty states? The answer is, by giving that electron some energy. There are various ways of giving an electron some energy, and we'll choose the easiest: heat. Let's suppose that our solid sodium is warmed up a little. The lone electron absorbs some heat and therefore gets more energetic. Now, it is, by definition, in a particular energy state or level because of the amount of energy it has. So if it obtains *more* energy, our lone electron will have to move up into a *higher* energy state. If the electron absorbs a bit more energy, it moves up again. And so on. In fact, our lone electron can move very easily up through the band because *there's nothing in its way*. It needs to absorb only a very small amount of energy to move from one state to the one above in the band because the states are so close together. There are no other electrons in there, so the electron can occupy any available energy state in the $n = 3$ and $n = 4$ bands. The electron can move easily through the band $n = 3$ and through $n = 4$ because these two bands overlap. The electron is then 'free' to wander about between the empty energy states in adjacent atoms.

Imagine, then, that our lone electron arrives in a higher energy state, possibly in the higher overlapping band, because it's gained energy. This electron is now available to take part in electrical conduction. If an electric field is applied across the sodium material using a battery, the free electron will wander off from empty state to empty state towards the positive pole. In other words, it forms a current. Imagine this happening in every atom of a lump of sodium; I hope you can see that if every atom produces one free electron there will be an awful lot of free electrons contributing to the current, which could be large enough to measure using an ordinary ammeter. In sodium the outermost electron is the 3s electron. This therefore occupies the band for which $n = 3$, and is the only electron in the shell. Hence the outermost band in solid sodium contains one 3s electron for every atom present. The atomic weight of sodium is 22.99 g mol^{-1}. One mole contains Avogadro's number ($N_A = 6.02 \times 10^{23}$) of atoms, so in 22.99 g of sodium there are 6.02×10^{23} 3s electrons. Hence sodium is a good conductor of electricity.

Before we move off the subject of sodium, let's consider the other electrons in a sodium atom occupying bands 1 and 2. Can they become free to move through the material? All these electrons occupy full shells: 2 per atom in band 1 and 8 in band 2. Because the shells are full there are no vacant energy states in those shells for an electron to move up into. They must gain enough energy to move up into the $n = 3$ shell (Figure 3.7). Hence an electron can't move up bit by bit in the same way that the lone electron in the third shell (n = 3) can. So the electrons in the inner, full shells are unlikely to contribute to a current.

SELF-ASSESSMENT QUESTION 3.1

How do free electrons form a current? Try to remember without referring to the above notes.

We've seen how the energy-band diagram of Figure 3.7 explains the good electrical conductivity of solid sodium. Energy-band diagrams can be used to classify materials, and later we'll use them to classify conductors, insulators and semiconductors.

3.1.2 Conduction and valence bands

The separation between the discrete levels is so small that the band of sublevels can be considered to be continuous. As with isolated atoms, the lowest-energy sublevel is occupied first, then the next highest, and so on, until all the electrons are accommodated.

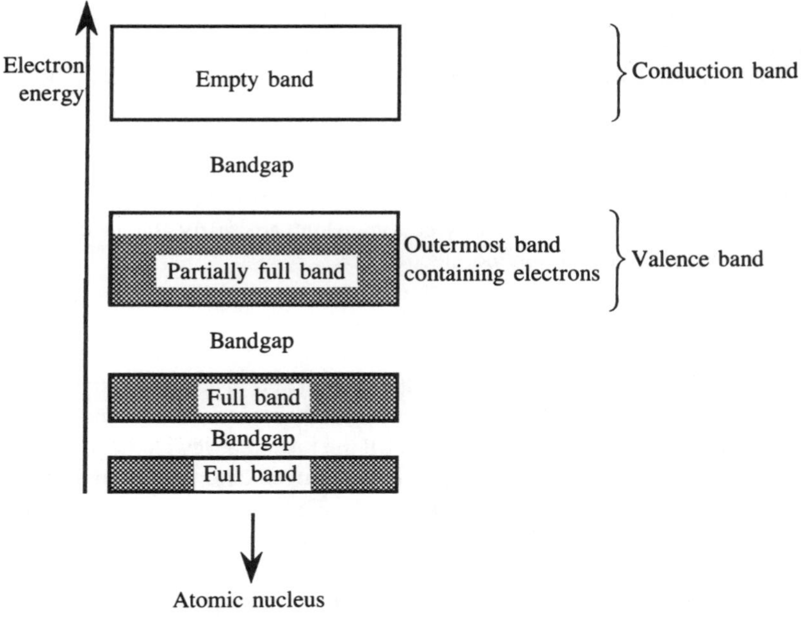

Figure 3.8 *Conduction and valence bands.*

Consider a solid in which there are a number of non-overlapping energy bands. Energy gaps (called bandgaps) exist between them (Figure 3.8). The electrons in the solid will occupy the lowest available energy states (or levels). The outermost band containing electrons is referred to as the valence band, and the electrons in this band are called valence electrons.

If an electron could, by some means, be given sufficient energy to cross the bandgap from the valence band to a state in the next band it could take part in conduction because it would be free to move from atom to atom. For this reason the first empty band above the highest filled band is called the conduction band.

3.2 Classification of solids

In this Section you'll find out how conductors, insulators and semiconductors can be classified according to their energy-band structure.

3.2.1 Conductors

If the outermost occupied band of the solid is not completely filled by electrons, it will be possible for the solid to conduct electricity in an applied electric field. For example, in the alkali metals such as sodium there is a single valence electron per atom in the outer subshell, whereas there are two states available for electrons to occupy in this subshell. Higher bands overlap such that even more empty states are available for conduction. Applying an electric field across the sample will then cause these free electrons to form a current, i.e. conduction will occur.

Generally, conductors always contain a partly filled band (or a partly filled set of overlapping bands), and the occupation of the available energy states is easily altered to produce a current by the application of an electric field. Conductors include the alkali metals (e.g. sodium, potassium) and metals (e.g. aluminium, mercury).

Metals are particularly useful because of their superb conducting properties. They contain a very large number of free electrons which are easily able to form a current under the influence of an applied electric field.

3.2.2 Insulators

In insulators at temperatures up to several hundred kelvin all bands are either almost completely full (of electrons) or almost completely empty. In many materials, there are just sufficient electrons to completely fill the lower energy bands while the upper energy bands remain completely empty. If, now, an electric field is applied to one of these materials with completely filled bands, the electrons cannot be accelerated in the direction of the field (i.e. a current cannot flow), since there are no available energy states to which the electron energies may be increased. There can thus be no conduction of electricity and the material will be an insulator. Note that the force exerted by a typical electric field is far too small to give an electron in an insulator the

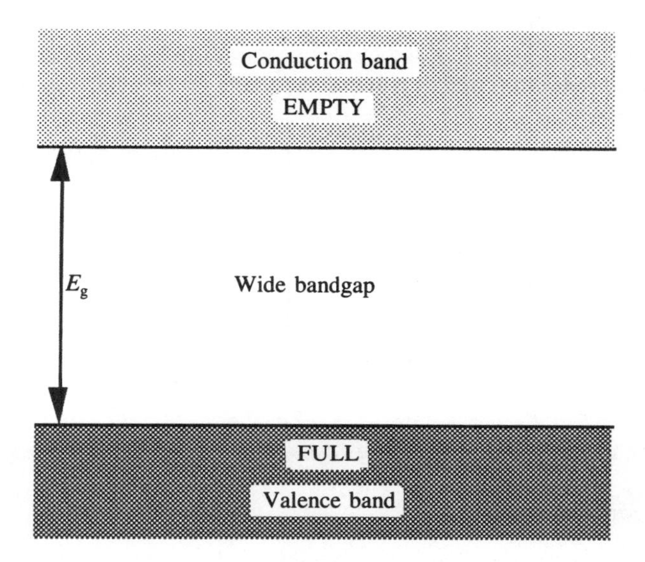

Figure 3.9 *Energy-band diagram for an insulator.*

energy required to cross the gap from the valence band to the conduction band.

The outermost band that contains electrons is completely full and is the valence band (Figure 3.9). The next band out is empty and is therefore the conduction band. The bandgap E_g separating these two bands is wide enough to prevent electrons being excited from the valence band to the conduction band under normal conditions, hence insulators don't conduct.

Dielectric breakdown

Normally a dielectric or insulator is a very poor conductor of electricity because the bandgap is too wide for electrons to be excited at normal temperatures and voltages. This means that applying energy to the material does not result in the formation of a current. However, if the energy is supplied in the form of a very large applied voltage V electrons will be ripped out of the valence band into the conduction band causing conduction. We can draw an I–V characteristic of the dielectric material. At low applied voltages there will be very little current arising from the very small number of electrons which manage to get into the conduction band. As the applied voltage is increased there is no increase in current until the voltage V exceeds a critical value V_b, as shown in Figure 3.10. A sudden increase in current occurs; this is dielectric breakdown. The typical time for dielectric breakdown in an insulator is 10 ns.

Examples of insulators include polyethylene, rubber and silica. Diamond is traditionally regarded as an insulator, but it is now possible to produce diamond that is semiconducting.

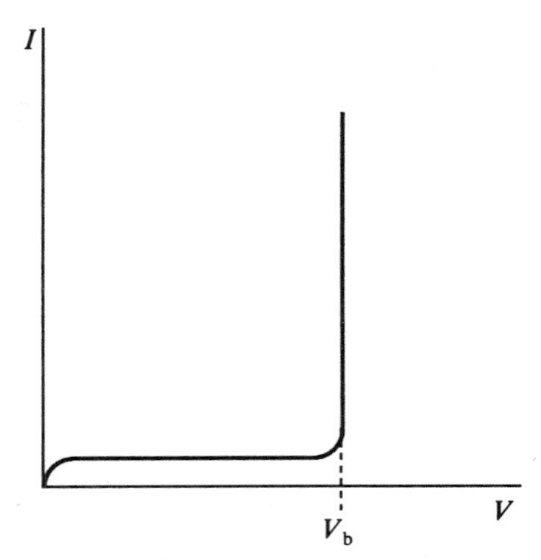

Figure 3.10 *I–V characteristic of a dielectric material, showing breakdown.*

3.2.3 Semiconductors

These have an energy-band structure similar to that of insulators, except that the bandgap E_g is narrow (Figure 3.11). The narrowness of the bandgap is important because some electrons can gain enough heat energy to move from the valence band to the conduction band at the sorts of temperature at which we use electronic equipment (i.e. room temperature or thereabouts). Note that relatively few valence electrons will gain enough energy to make this transition, so that the conductivity of semiconductors, while greater than that for insulators, is a lot lower than that for metals.

At temperature $T = 0$ K

Semiconductors at temperatures near 0 K (or −273.15°C) are virtually the same as insulators in that they do not conduct. This is because the temperature is so low that it's

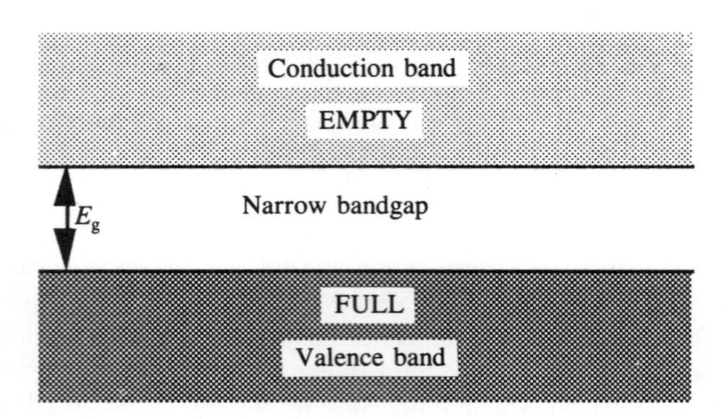

Figure 3.11 *Energy-band diagram for a semiconductor at T=0 K.*

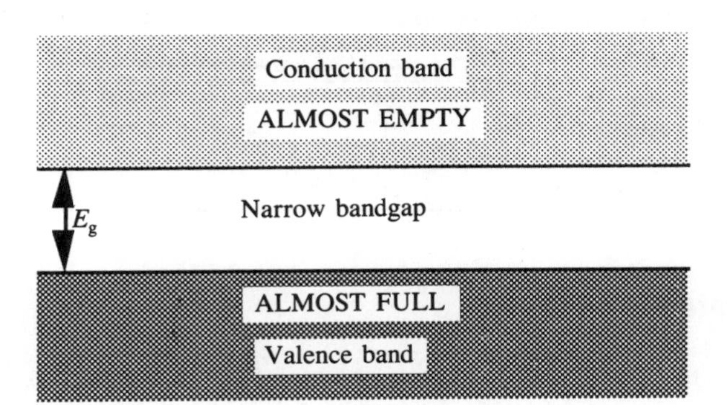

Figure 3.12 *Energy-band diagram for a semiconductor at $T > 0$ K.*

impossible to give them enough thermal energy to break their bonds, so they stay firmly stuck in the valence band. So electrons cannot cross the bandgap into the conduction band (Figure 3.11).

At temperature $T > 0$ K

Semiconductors will conduct at room temperature. This is due to the thermal excitation of electrons from the valence band into the conduction band. This is possible only because the bandgap is narrow (Figure 3.12). Once in the conduction band the electrons are available for conduction. Many more electrons are excited into the conduction band at room temperature (300 K) than at lower temperatures.

So the valence-band electrons may get enough thermal energy to be excited across the bandgap into available states. The valence band is now only partly empty because some valence electrons have been excited into the conduction band. Of course this means that the conduction band is now partly full as it contains the excited valence electrons. Remember that the excited electrons are not allowed to occupy the bandgap — they either get enough energy to cross the bandgap into the conduction band, or they stay in the valence band.

3.2.4 **Bandgaps of some materials**

Table 3.1 lists some bandgaps of common materials. Germanium, silicon and gallium arsenide are common semiconductors. Silicon dioxide is a glassy material which is commonly used in electronic devices.

SELF-ASSESSMENT QUESTION 3.2

Look at Table 3.1, listing bandgaps of some materials. One of them is used as an insulator. Which one do you suppose it is, and why?

Table 3.1 *Bandgaps of some common electronic materials*

Material	*Bandgap (eV) at room temperature*
Germanium (Ge)	0.6
Silicon (Si)	1.1
Gallium arsenide (GaAs)	1.4
Silicon dioxide (SiO_2)	9.0

3.3 Crystals

The way atoms form solid materials is important in electronics because the environment the electron occupies determines its behaviour. A particularly important form of solid is the crystal. Many elements and compounds are crystalline, including the common semiconductors.

In structure, a crystal is a highly ordered, three-dimensional grid which contains atoms at the grid intersections. The lines formed by the grid are called the lattice. The atoms lie in flat planes, and the planes are stacked to form the crystal. The crystal structure of the common semiconductors is based on one type of lattice: the cubic lattice, which will be described here.

The simplest form of cubic lattice is shown in Figure 3.13 — it's known as the simple-cubic (SC) crystal structure. One layer of atoms (called a plane) is shown in Figure 3.13(a), and a three-dimensional representations showing a stack of these planes is shown in Figure 3.13(b). The smallest group of atoms which represent the crystal is called a unit cell; when repeated throughout space (i.e. along all three dimensions) the unit cell replicates the whole of the crystal. Figure 3.13(c) shows the unit cell for the SC crystal structure.

There is one atom at each of the eight corners of the SC unit cell, shown by the dark spheres in Figure 3.13(b). If you look carefully at Figure 3.14 you should be able to see that each atom is shared between eight cells, so *within* the unit cell itself there is one-

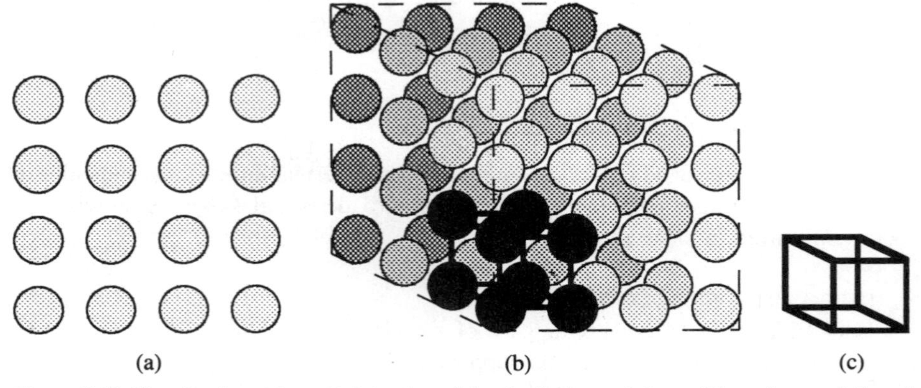

Figure 3.13 ***The simple-cubic crystal structure: (a) a single plane of atoms; (b) a cube consisting of a stack of four planes, showing a cubic unit cell in one corner; (c) a unit cell.***

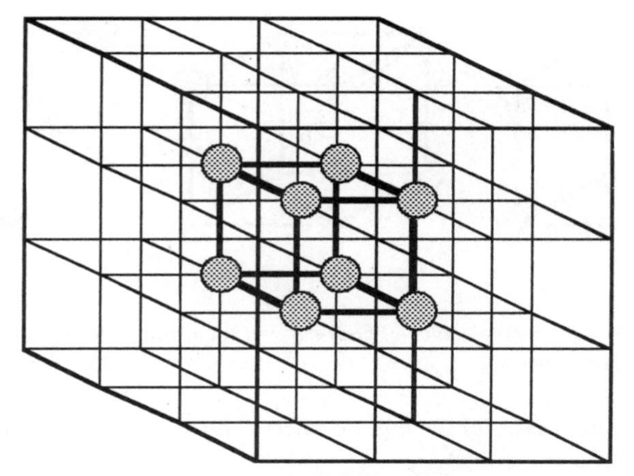

Figure 3.14 *The simple-cubic unit cell, showing how it shares eight atoms with its neighbouring cells.*

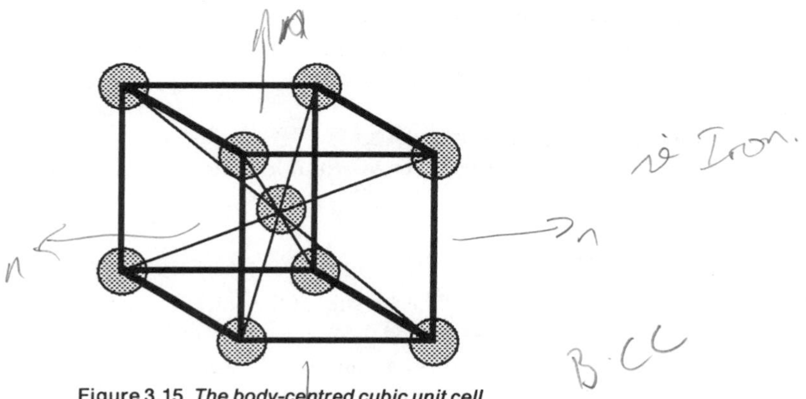

Figure 3.15 *The body-centred cubic unit cell.*

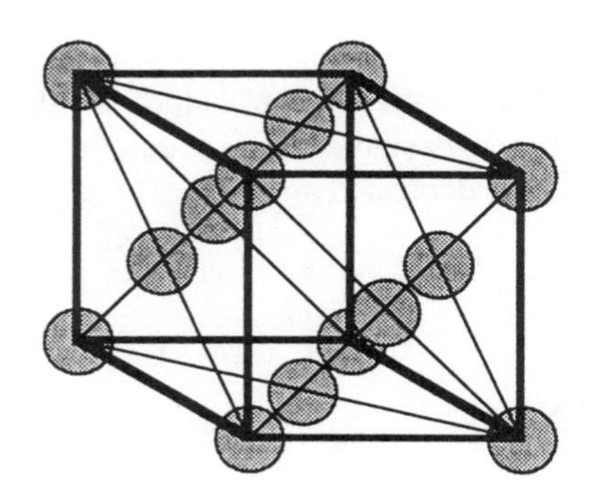

Figure 3.16 *The face-centred cubic unit cell.*

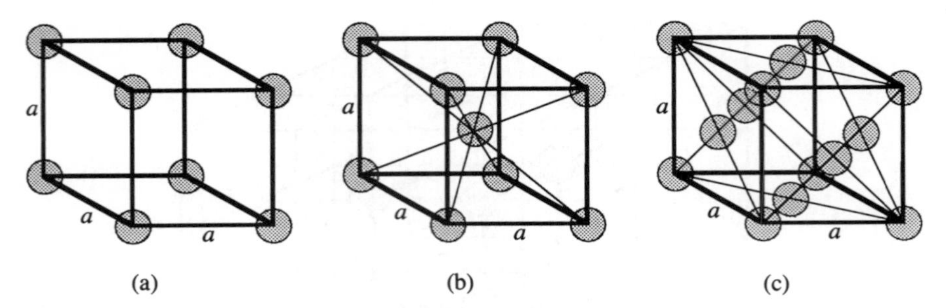

Figure 3.17 *Cells of (a) simple-cubic, (b) body-centred cubic and (c) face-centred cubic crystals, showing the lattice constant* a.

eighth of an atom at each of the eight corners. The simple-cubic unit cell therefore contains one atom.

There are two other types of cubic crystal: the body-centred cubic (BCC) crystal and the face-centred cubic (FCC) crystal. In the BCC structure (Figure 3.15) there is one atom at each of the eight corners and one atom in the centre of the cube. In the FCC structure (Figure 3.16) there is one atom at each corner of the cell, plus one atom in the centre of each of the six faces.

SELF-ASSESSMENT QUESTION 3.3

How many atoms are contained within one unit cell of the FCC structure?

3.3.1 **Lattice constant**

In cubic crystals the length of each side of the cell is the same. This is the quantity a shown in Figure 3.17 and is called the lattice constant. Table 3.2 lists lattice constants for several cubic crystals. a is measured in Å. Note that the lattice constant a is not necessarily the smallest distance separating atoms in a cell.

You can see from Table 3.2 that some materials have very similar, or even identical, lattice constants.

RESEARCH 3.1

Cadmium selenide and indium arsenide have identical lattice constants, 6.04 Å. Do both these compounds have the same crystal structure? If they have, it would mean that layers of CdSe could be deposited on layers of InAs without any distortion in the lattice.

3.3.2 **The diamond lattice**

Most common semiconductors have a very particular lattice structure which consists of two interlocking face-centred-cubic lattices (Figure 3.18). The unit cell looks like an FCC cell which contains four extra atoms, making a total of eight atoms. The structure can most easily be seen by breaking down the unit cell into a pattern of smaller cells, each of which contains five atoms in the form of a tetrahedron (Figure 3.18(a)). When all atoms in the cells are the same atomic spheres (Figure 3.18(b)), the structure is named after diamond, in which all the atoms are carbon. Examples of semiconductors that have a diamond structure are silicon and germanium.

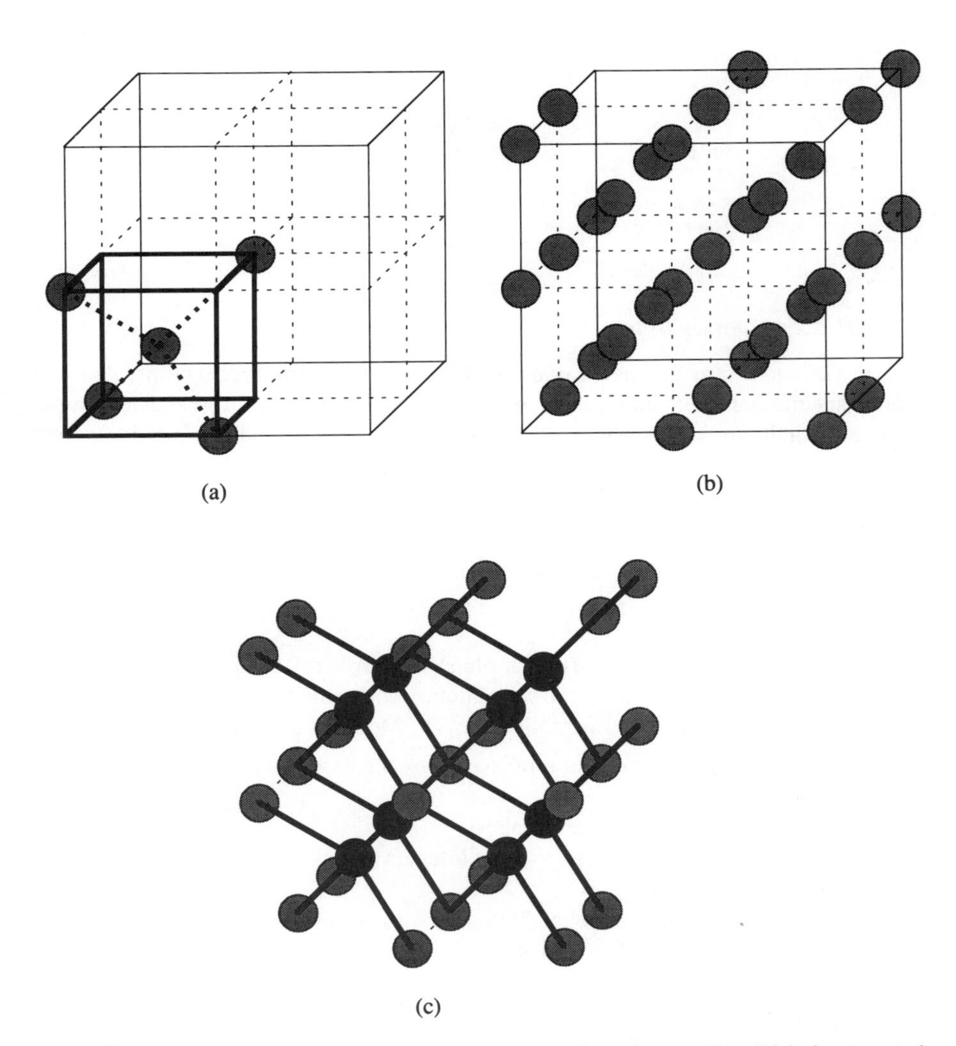

Figure 3.18 *Semiconductor lattices: (a) the basic tetrahedral structure which is repeated throughout; (b) the diamond lattice; (c) the zincblende lattice, showing alternate layers of two atomic species.*

Table 3.2 *Lattice constant* a *for various solid materials*

Material		*a(Å)*
C	diamond	3.56
Cu	copper	3.61
Na	sodium	4.29
FeO	iron oxide	4.29
Si	silicon	5.42
Ge	germanium	5.62
AlAs	aluminium arsenide	5.63
GaAs	gallium arsenide	5.64
InP	indium phosphide	5.86
CdSe	cadmium selenide	6.04
InAs	indium arsenide	6.04
PbTe	lead telluride	6.44

Source: Lide, D.R. (ed.), *CRC Handbook of Chemistry and Physics*, 75th edn, CRC Press, 1994.

3.3.3 **The zincblende lattice**

This has the same structure as the diamond lattice, but this time there is more than one atomic species (i.e. more than one element) in the lattice (Figure 3.18(c)). An example of a semiconductor which has a zincblende lattice is GaAs, which is a compound semiconductor made up of gallium, Ga, atoms and arsenic, As, atoms. In GaAs, half the atoms are Ga, and half are As. In Figure 3.18(c) you can seen the alternating layers of the two atomic species.

3.3.4 **Crystal planes**

It's useful to be able to refer to particular planes within a crystal lattice. Three integers are used called Miller indices, where each integer represents one of the coordinates x, y or z. The integers that are used are called hkl. The Miller indices are found as follows:

1 Find the intercepts on the axes represented by the basis vectors **a**, **b**, **c** in terms of the lattice constant a.
2 Take the reciprocals of these numbers.
3 Reduce these three reciprocals to three integers having the same ratio, usually the smallest three integers.
4 These three new integers are the Miller indices, represented by (hkl). The round brackets indicate that hkl denotes a plane.

EXAMPLE 3.1

Figure 3.19 shows part of a plane (the shaded triangle) which intersects the three axes x, y, z of a coordinate system. The three axes show the basis vectors **a**, **b**, and **c**.

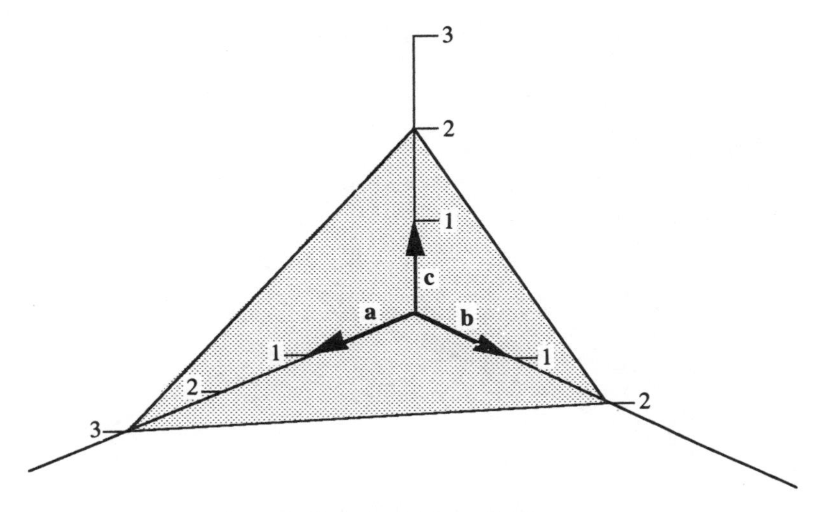

Figure 3.19 *Figure for Example 3.1.*

Apply the rules for determining the Miller indices:

1 The intercepts are 3**a**, 2**b** and 2**c**.
2 The reciprocals are $\frac{1}{3}$, $\frac{1}{2}$, $\frac{1}{2}$.
3 The smallest three integers having the same ratio are 2, 3, 3 (each reciprocal has been multiplied by 6).
4 The Miller indices denoting this plane are therefore (233). This plane is called the (233) plane.

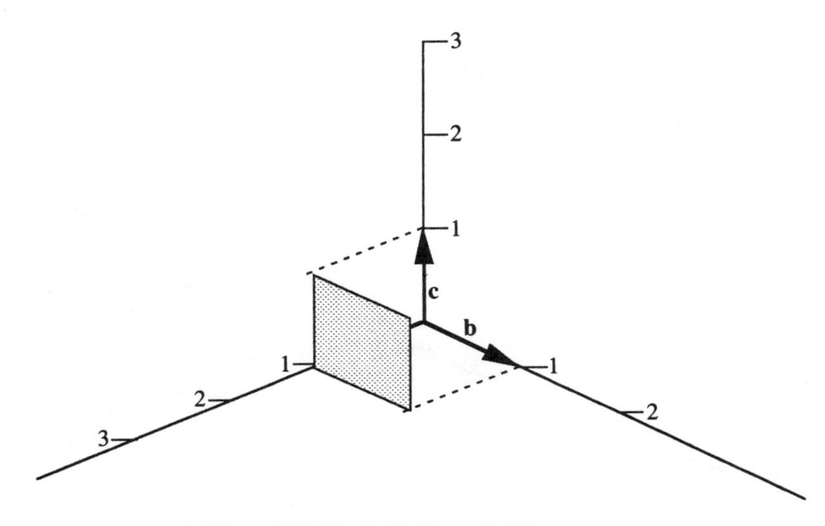

Figure 3.20 *Figure for Example 3.2.*

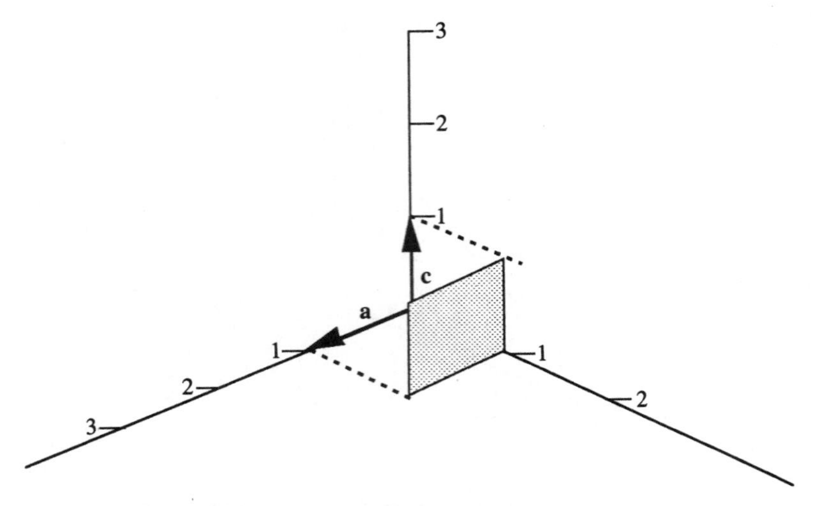

Figure 3.21 *Figure for Self-assessment question 3.4.*

EXAMPLE 3.2

The plane of Example 3.1 is rather an obscure example for us to consider. Semiconductor engineers are usually more concerned with the planes that make up a simple-cubic lattice, and one of these simple-cubic planes is used in this example. Consider the plane of Figure 3.20. Note that it intersects only one axis. Now apply the rules:

1 The intercepts are 1**a**, ∞**b** and ∞**c**.
2 The reciprocals are 1, 0, 0.

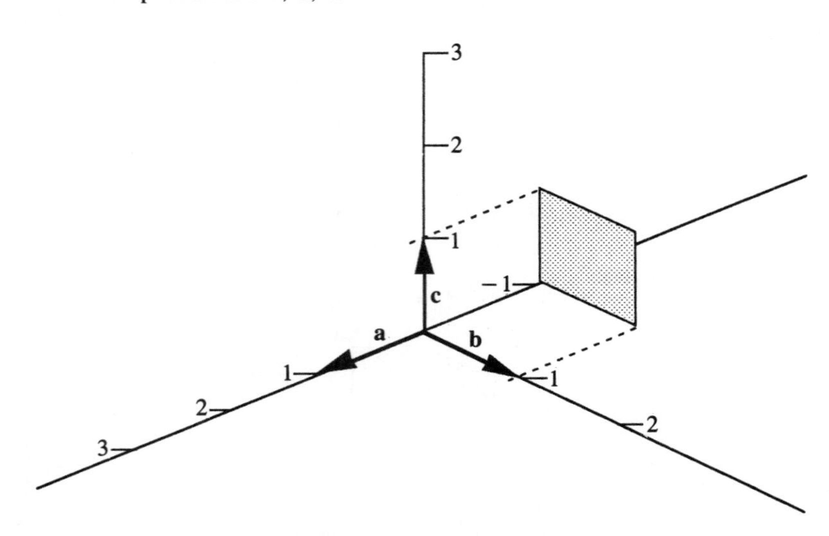

Figure 3.22 *Figure for Self-assessment question 3.5.*

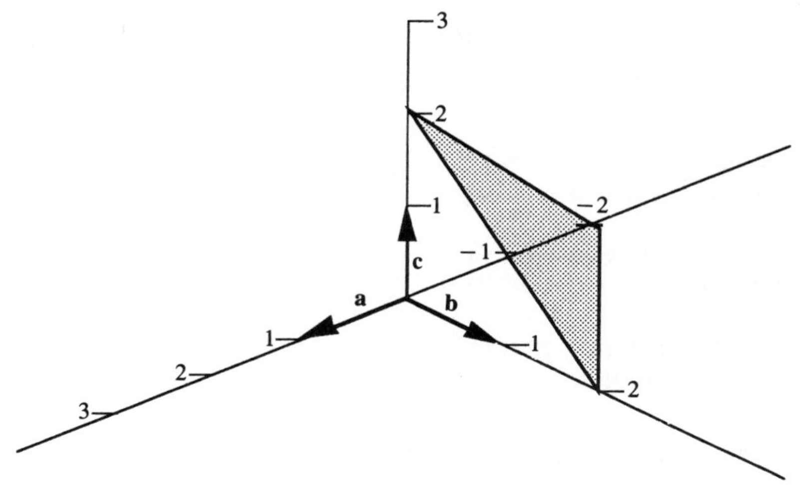

Figure 3.23 *Figure for Self-assessment question 3.6.*

3 The smallest three integers having the same ratio are 1, 0, 0.
4 The Miller indices are therefore (100). This plane is called the (100) plane.

SELF-ASSESSMENT QUESTION 3.4
Now try one for yourself. What is the plane shown in Figure 3.21?

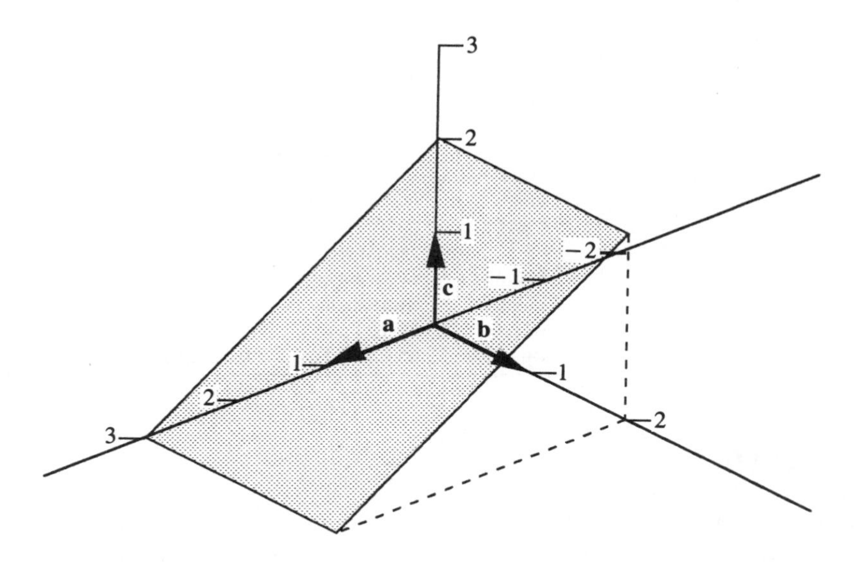

Figure 3.24 *Figure for Self-assessment question 3.7.*

Cubic planes

The faces of a cubic crystal are the (100), (010), (001), $(\bar{1}00)$, $(0\bar{1}0)$ and $(00\bar{1})$ planes. The minus sign (often called a bar) above the index shows that the plane has cut the axis on the negative side of the origin.

SELF-ASSESSMENT QUESTION 3.5
Now try another one (Figure 3.22). It's another cubic plane, but which one?

SELF-ASSESSMENT QUESTION 3.6
Identify the plane in Figure 3.23.

SELF-ASSESSMENT QUESTION 3.7
Identify the plane in Figure 3.24.

3.3.5 *Elemental and compound semiconductors*

Table 3.3 gives some examples of elemental and compound semiconductors. The compound semiconductors are either binary (two atomic species), ternary (three atomic species), or quaternary (four atomic species). The relevant group from the Periodic Table is also shown, as this is a common way of naming semiconductor types. For example, A1As and GaInAsP are both III–V compound semiconductors. The elemental semiconductors listed have a diamond lattice, and the compound semiconductors listed all have a zincblende lattice.

Table 3.3 *Classification of some common semiconductors according to Periodic-Table group*

Type of semiconductor		*Semiconductor*		*Group(s)*
Elemental		C	diamond	IV
		Si	silicon	IV
		Ge	germanium	IV
Compound	Binary	CdTe	cadmium telluride	II–VI
		ZnSe	zinc selenide	II–VI
		AlAs	aluminium arsenide	III-V
		GaAs	gallium arsenide	III–V
		InP	indium phosphide	III–V
		SiC	silicon carbide	IV
		PbTe	lead telluride	IV–VI
	Ternary	AlGaAs	aluminium gallium arsenide	III–V
	Quaternary	GaInAsP	gallium indium arsenide phosphide	III–V

3.4 Summary

As atoms move closer together as they change from gas to liquid to solid the outer energy levels split to form energy bands. Electrons are easily able to move up through an energy band if they absorb energy. The energy bands may overlap each other and may extend into the continuum of positive energies representing electrons which have left the material completely. An electron may absorb enough energy for it to move up through the band, becoming a free electron. If an electric field is applied to a material containing free electrons those free electrons will form a current.

Energy bands explain why some materials, such as sodium, are good conductors of electricity. Currents can flow only in solids which contain a band partly filled with electrons, or an empty band close to or overlapping a lower band containing electrons. The electrons in a completely filled band cannot carry current because there are no available energy states in there for electrons to occupy.

Conductors have no bandgap above the outermost band containing electrons, therefore there are many empty states for conducting electrons to occupy. Hence a lot of electrons can be excited into these many empty states so that large currents can be obtained. Insulators have a wide bandgap above the outermost full band, making it very difficult for an electron to be excited out of the full band and into the next empty band. Dielectric breakdown is the phenomenon which occurs when enough electrical energy is applied to insulator electrons to cause them to break their bonds and move up into the empty band. Semiconductors have a narrow bandgap above a full band so that electrons are easily able to gain enough thermal energy to break their bonds and be excited up into the empty band. Hence semiconductors can form a current.

The common semiconductors are crystalline, having a cubic lattice. Elemental semiconductors have a diamond lattice, whereas compound semiconductors have a zincblende lattice. Miller indices are used to identify planes of atoms within crystals. Elemental and compound semiconductors can be classified according to the Periodic-Table groups of their constituent atoms.

3.5 **Tutorial questions**

3.1 In your own words, explain why sodium is a good conductor of electricity.

3.2 Would you expect semiconductor conductivity to increase or decrease with temperature? Explain your answer.

3.3 Would you expect metal conductivity to increase or decrease with temperature? Explain your answer.

3.4 Determine the smallest distances between atoms in a simple-cubic, a body-centred cubic and a face-centred cubic cell. Express your answers in terms of the lattice constant a.

3.5 Assuming that the crystal structure of sodium is body-centred cubic, calculate the density of sodium. Assume the following information:

Atomic weight of sodium, Na = 22.99
Sodium lattice constant a = 4.3 Å

3.6 Calculate the density of indium arsenide, InAs, assuming it has the zinc-blende crystal structure. Also assume the following information:

Atomic weight of indium, In = 114.82

Atomic weight of arsenic, As = 74.9
Lattice constant of InAs = 6.04 Å

3.7 Determine the number of indium atoms per square centimetre on the (100) surface of an InAs wafer.

3.8 Assuming that there are eight atoms in a unit cell of silicon, calculate the density of silicon. Assume the following information:

Atomic weight of silicon = 28.09
Silicon lattice constant a = 5.43 Å

3.6 Suggested further reading

Anderson, J.C., Leaver, K.D., Rawlings, R.D. and Alexander, J.M., *Materials Science*, 3rd edn, Van Nostrand Reinhold, 1985, pp. 70–84, 88–104.

CHAPTER 4

Semiconductors and carrier transport

Aims and objectives

This chapter describes intrinsic and n-type and p-type extrinsic semiconductor materials, and describes simple physical models for their electrical behaviour. A two-dimensional physical picture is given for the movement of holes and electrons through semiconductor lattices, as well as simple band-structure models. Hole conduction is explained using the band-structure model applied to an intrinsic material and the concept of majority and minority carriers is introduced when the model is applied to extrinsic semiconductor materials. The idea of the Fermi level is introduced as a means of classifying semiconductors. Work function and electron affinity are explained using the band-structure model. Fundamental equations are described and derived for the carrier-transport mechanisms drift and diffusion, starting off with a brief description of carrier motion in a crystalline solid. Relationships are found between many important electrical device parameters such as resistance, resistivity, conductivity, mobility, majority-carrier concentration, drift velocity and diffusion coefficient. The chapter ends by examining the Hall effect to determine carrier type, majority-carrier concentration and mobility.

4.1 Classification of semiconductors

p-type and n-type semiconductors are both extrinsic semiconductor materials. An extrinsic semiconductor is obtained by adding to a pure sample (called an intrinsic semiconductor) certain impurities to make it a better conductor of electricity. Once these impurities have been purposely added, the material becomes extrinsic. The type of impurity which is added determines whether the semiconductor will be n-type or p-type.

4.1.1 Intrinsic semiconductors

This is the so-called 'pure' state, but note that, in practice, intrinsic materials may contain a very small number of impurity atoms which aren't really supposed to be there. These unwanted impurity atoms may be incorporated into the semiconductor lattice as a result of factory-processing methods. Ideally then intrinsic semiconductors contain only atoms of the semiconductor itself. These atoms bond together to form a solid. The common elemental semiconductors (e.g. silicon Si, germanium Ge) have a valency of four, i.e. they are tetravalent and they have four valence electrons. Compound semiconductors such as gallium arsenide GaAs and indium phosphide InP have an 'average' valency of four because they consist of a group III atom (Ga, In) and a group V atom (As, P). The bonding in all these semiconductors is, generally, covalent, such that the atoms hold together by 'sharing' their valence electrons with their four nearest neighbours. We can draw a simple picture to show how silicon atoms bond to form the solid (Figure 4.1).

These covalent bonds are very strong, and the crystal structure is very stable. At absolute zero, the lattice would be perfectly still and the semiconductor would behave like an insulator: all the valence electrons remain in their bonds. Figure 4.2 shows how

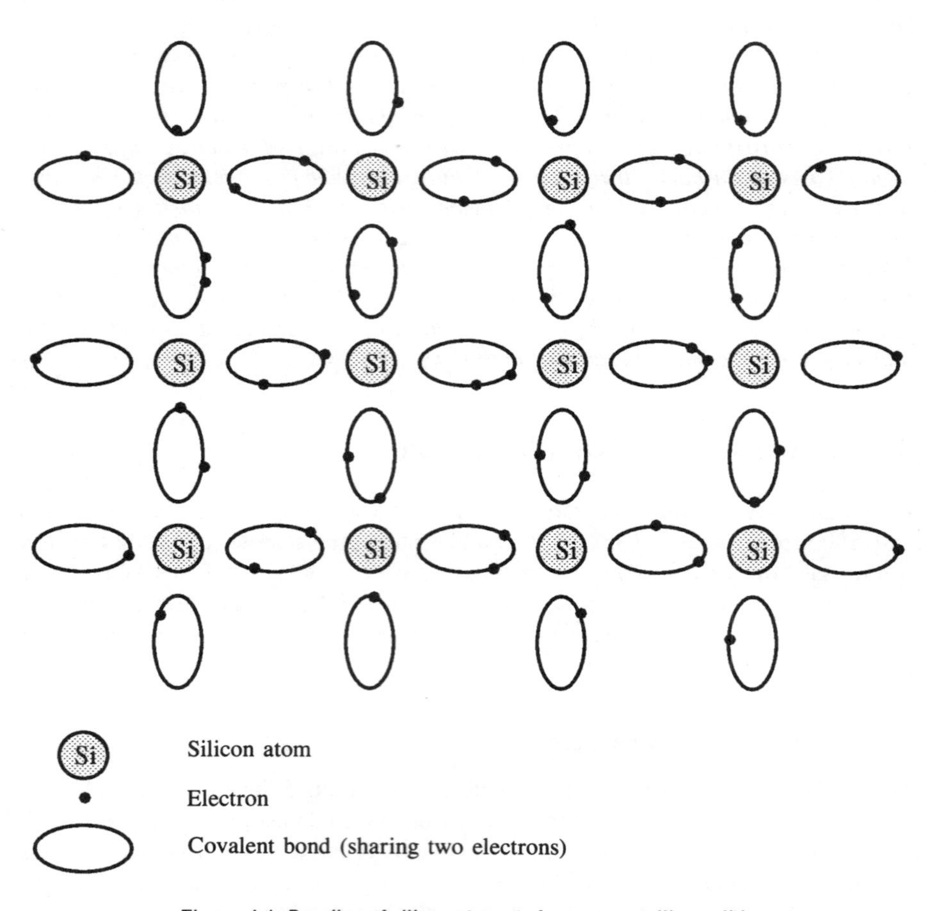

Figure 4.1 *Bonding of silicon atoms to form a crystalline solid.*

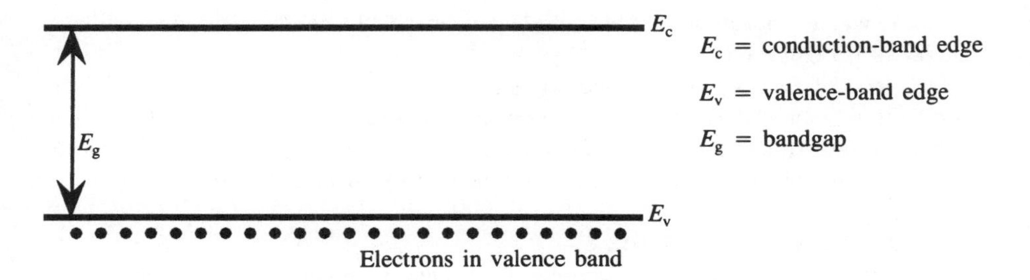

Figure 4.2 *Energy-band diagram for an intrinsic semiconductor at $T = 0$ K.*

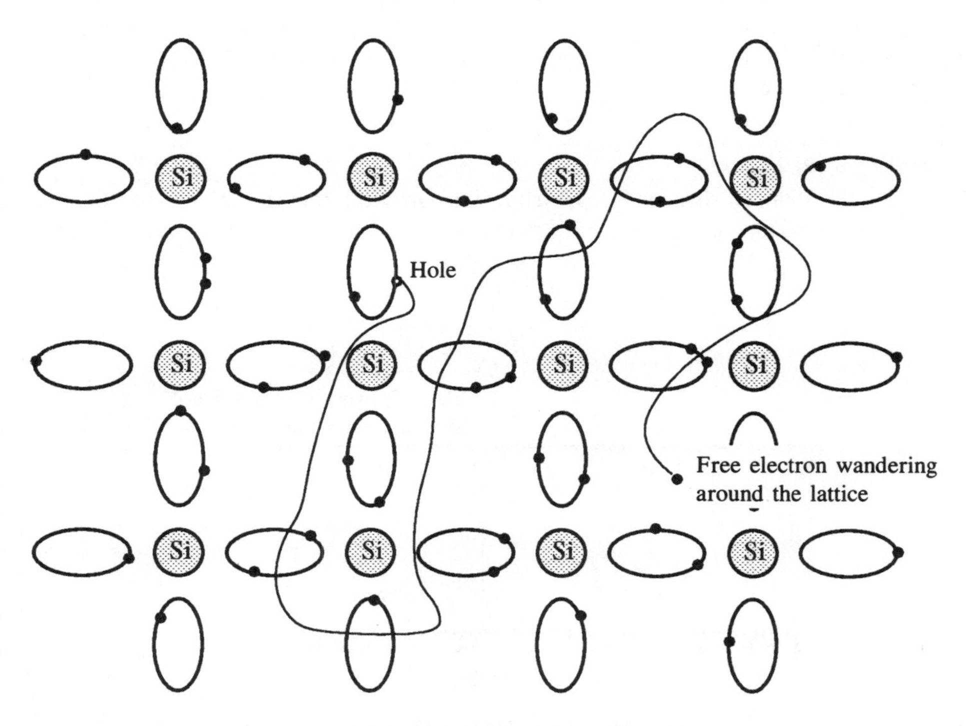

Figure 4.3 *Intrinsic silicon lattice at $T > 0$ K.*

an intrinsic semiconductor at absolute zero would be represented by an energy-band diagram.

As the temperature increases a valence electron may somehow receive enough energy to leave its bond. This valence electron is then called a 'free' electron because it is free to wander the lattice. The electron leaves a gap in the bond — this gap is called a 'hole' (see Figure 4.3).

In terms of energy, the free electron is now located in the conduction band. The remaining valence electrons are located in the valence band. Figure 4.4 shows that there is a hole in the valence band corresponding to the electron in the conduction band. Together, this electron and the hole are called an electron–hole pair.

SELF-ASSESSMENT QUESTION 4.1

Consider a completely pure sample of intrinsic germanium. Suppose that, at a particular temperature, the density of electrons in the conduction band is 1×10^{12} cm^{-3}. What will the hole density be in the valence band at the same temperature, and why?

SELF-ASSESSMENT QUESTION 4.2

Consider the completely pure sample of intrinsic germanium again (see Self-assessment question 4.1). Suppose that the temperature changes. Would you expect the density of electrons in the conduction band to change? Would the hole density in the valence band still equal the electron density in the conduction band?

Hole conduction

At $T > 0$ K then, an electron has somehow been excited from the valence band to the conduction band, leaving a hole in the valence band. What happens if an electric field $\mathcal{E}$

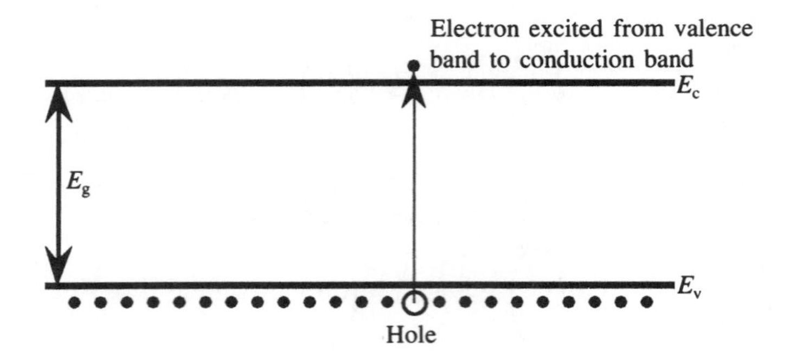

Figure 4.4 *Energy-band diagram for an intrinsic semiconductor at $T > 0$ K.*

is applied across the material? Say the field is applied from right to left, such that the right-hand side is positive relative to the left-hand side (Figure 4.5(a)).

What happens next? The electron in the conduction band is attracted towards the positive side of the lattice, and therefore moves, i.e. it causes conduction in the conduction band (Figure 4.5(b)). Now consider the hole in the valence band. The electron to the left of the hole would like to move towards the positive side. It can do this if it occupies the hole left by the original electron. So the electron moves into the hole, creating a new hole in the process. As in the conduction band, the electron moves

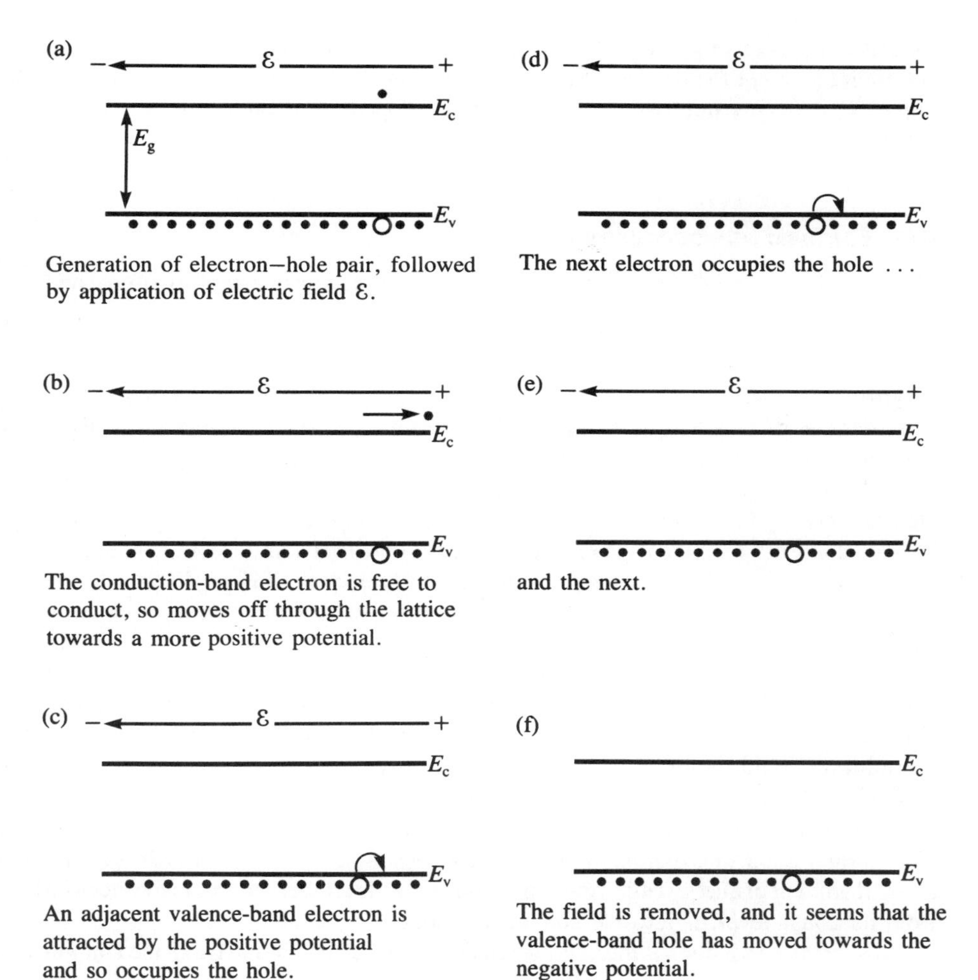

Figure 4.5 *Energy-band diagrams for an intrinsic semiconductor at $T > 0$ K under the influence of an applied electric field ℰ, showing how hole conduction occurs.*

from left to right (Figure 4.5(c)). The field $\mathscr{E}$ is still applied, so the electron to the left of the new hole moves into that hole, leaving behind yet another new hole (Figure 4.5(d)). The process continues as long as the field is applied (Figure 4.5(e)). When the field is removed (Figure 4.5(f)), we see that the conduction-band electron has gone elsewhere in the lattice while conducting. In the valence band, however, an electron is seen to have moved from left to right. *In effect, the hole has moved from right to left*. Hence we can say there are two modes of conduction:

1 electron conduction in the conduction band;
2 hole conduction in the valence band.

Intrinsic semiconductors therefore do conduct, but there are not many excited electrons at normal temperatures so conduction in minimal, and any currents flowing are often too small to be of use. Such materials may have to be doped with specific impurity atoms to improve conduction. Such doped semiconductors are called extrinsic.

4.1.2 **Extrinsic semiconductors**

There are two types, depending on the type of carrier (i.e. hole or electron): n-type and p-type.

n-type

Consider the silicon lattice as before, but this time remove a silicon atom and replace it with an antimony atom. Antimony (Sb) is pentavalent, i.e. it has five valence electrons. Four of its electrons form covalent bonds with the four nearest neighbours, but the fifth electron is only loosely bound to its parent antimony atom. This fifth electron forms a local unstable orbit around its parent atom (Figure 4.6(a)). If a small amount of energy enters the lattice it may be enough to free the fifth electron (Figure 4.6(b)). In the presence of an electric field, this freed electron would contribute towards an electron current in the conduction band. This has been achieved by inputting only a small amount of energy, possibly by the material just being at room temperature. Note that antimony is a large atom and could distort the lattice. Care has to be taken when doping with large atoms — if they're too large, the bonding could be broken and the crystallinity could be lost.

Figure 4.7 shows how the presence of the fifth 'dopant' electron contributes to conduction in terms of energy-band structure. Pentavalent atoms such as antimony are called donor impurities or donor atoms because they donate an electron to the system. A small amount of energy called the activation energy is required to excite the electron from its donor level E_d into the conduction band.

Most of the conduction is therefore by electrons in the conduction band; the amount of conduction depends on the number of donor atoms in the lattice. These electrons are referred to as the majority carriers because most of the conduction occurs because of them. There will be some free holes in the valence band because of the generation of

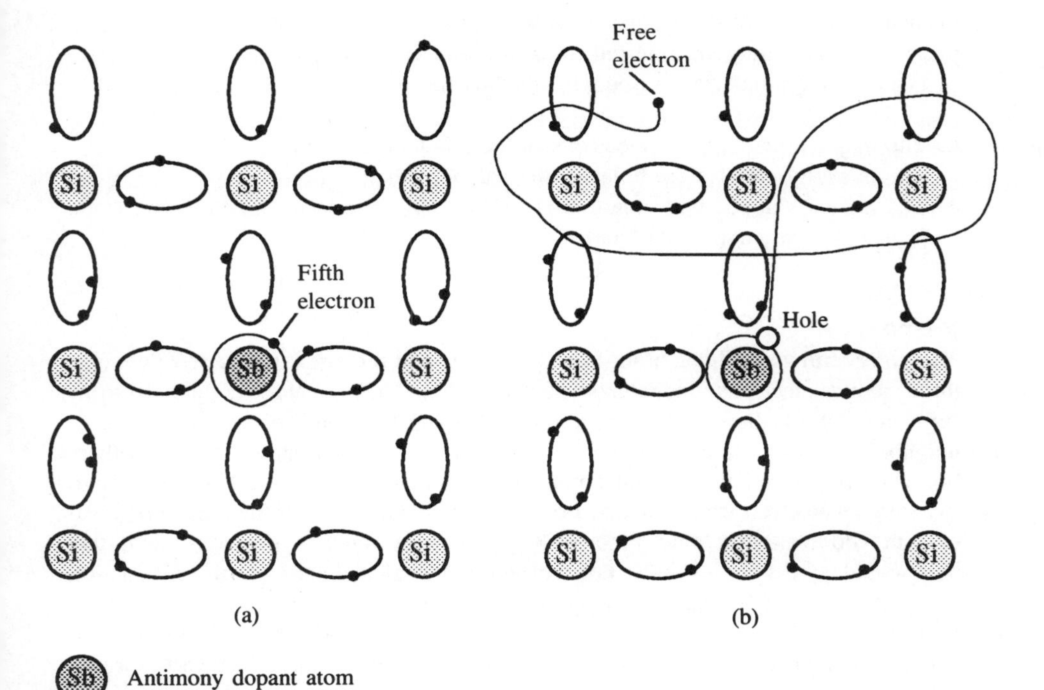

Figure 4.6 *Silicon lattice containing n-type dopant atom (antimony).*

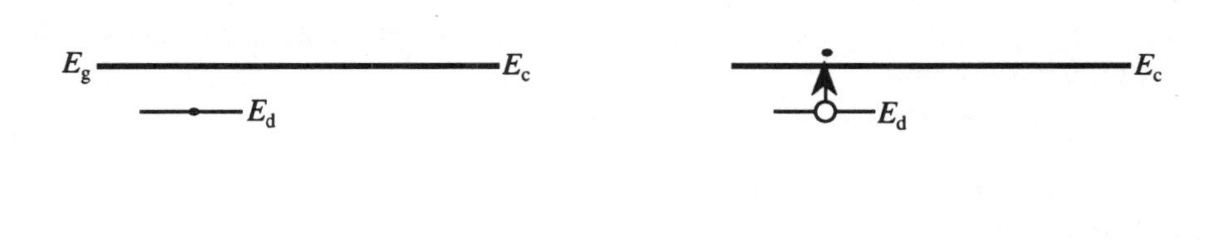

(a) The fifth electron occupies a localized donor level of energy E_d.
Note that, for convenience, the valence-band edge has been designated by energy $E = 0$.

(b) The donor electron is excited into the conduction band, where it is free to conduct.

Activation energy $= E_g - E_d$

Figure 4.7 *Energy-band diagram for n-type doping by a donor atom.*

intrinsic electron–hole pairs, and these will conduct. However, there aren't many of these free holes; they are referred to as minority carriers.

Hence n-type material can be described as follows:

- The majority carriers are electrons in the conduction band.
- The minority carriers are holes in the valence band.
- The lattice is doped with donor atoms which give rise to localized donor levels just below the conduction band.

p-type

This time remove a silicon atom from the lattice and replace it with a trivalent (i.e. three valence electrons) dopant atom such as indium (In). The three dopant electrons form bonds with three nearest neighbours, leaving the bond to the fourth nearest neighbour with a hole in it (Figure 4.8(a)). This bond is less stable than the others because it is incomplete. If a small amount of energy enters the lattice it may be enough to excite an electron from a nearby bond into the hole, such that the hole appears to have moved to another bond (Figure 4.8(b)). In the presence of an electric field, this hole would contribute towards a hole current in the valence band. Again, conduction has been achieved by inputting only a small amount of energy.

The energy-band diagrams describing p-type material are shown in Figure 4.9. Trivalent dopant atoms are called acceptor impurities because they create holes which accept electrons. Figure 4.9(a) shows the hole in the bond occupying a localized acceptor level E_a which is located just above the valence band.

A small amount of energy, the activation energy, is required to excite a valence electron into the hole (Figure 4.9(b)), leaving a hole in the valence band. This valence-band hole will then be available for conduction in the valence band. The size of the hole current achieved will depend on the number of acceptor atoms in the lattice. There will be some free electrons in the conduction band because of the generation of intrinsic electron–hole pairs, and these will conduct. However, there are very few free electrons compared to free holes and electron conductivity is correspondingly low.

Hence p-type material can be described as follows:

- The majority carriers are holes in the valence band.
- The minority carriers are electrons in the conduction band.
- The lattice is doped with acceptor atoms which give rise to localized acceptor levels just above the valence band.

SELF-ASSESSMENT QUESTION 4.3

Consider a sample of n-type silicon which has a bandgap of 1.1 eV. Assume it has been doped such that the dopant energy level E_d is 0.2 eV below the conduction-band edge E_c. What is the activation energy of this sample? (*Hint:* Sketch the energy-band diagram first — it will help you.)

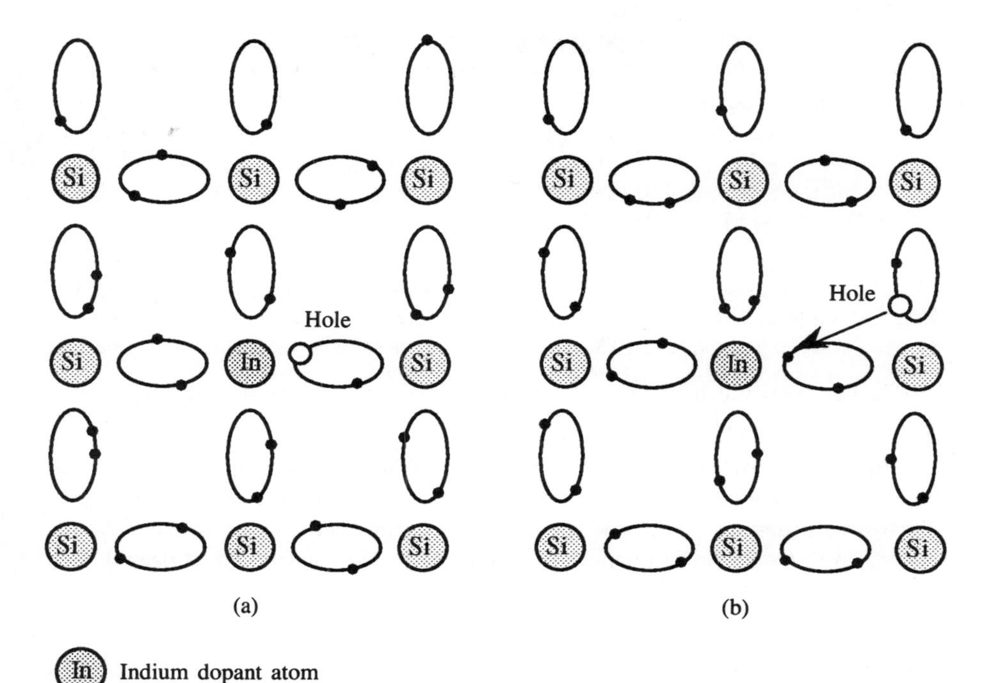

Figure 4.8 *Silicon lattice containing p-type dopant atom (indium).*

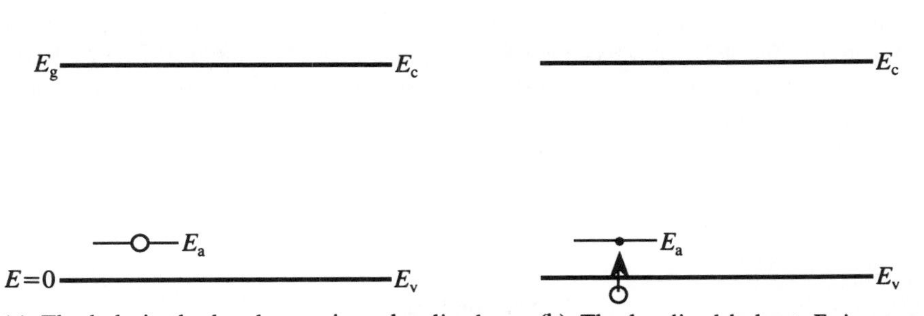

(a) The hole in the bond occupies a localized acceptor level of energy E_a.
Note that, for convenience, the valence-band edge has been designated by energy $E=0$.

(b) The localized hole at E_a is occupied by a neighbouring valence electron from the lattice. This leaves a hole in the valence band width which is available for conduction.

Activation energy $= E_a$.

Figure 4.9 *Energy-band diagram for p-type doping by an acceptor atom.*

SELF-ASSESSMENT QUESTION 4.4

Consider a sample of p-type gallium arsenide which has a bandgap of 1.4 eV. Assume it has been doped such that the dopant energy level E_a is 0.8 eV below the conduction-band edge E_c. What is the activation energy of this sample? (*Hint:* Sketch the energy-band diagram first — it will help you.)

4.2 The Fermi level

This is an imaginary energy state (or level) which is used as a reference point. Engineers often need to know how far an energy state is from another, or how far it is from a known reference point. This known reference is the Fermi level. It is defined as:

The Fermi level is that energy level for which there is a 50% probability of being occupied by an electron.

Why do we need to talk in terms of probability? This is because engineers deal with devices which contain very many free electrons, say 10^{20} cm^{-3}, and because all the electrons are identical there is no way of distinguishing between them, except by the energy levels (or states) they happen to occupy. A convenient way of dealing with such massive populations of identical items is statistically.

4.2.1 The Fermi level in intrinsic semiconductors

Consider an intrinsic material. Such a material is not doped, so the free electrons that are excited into the conduction band must have come from the valence band. So for every electron in the conduction band there must be a hole in the valence band. So the number of electrons is equal to the number of holes.The electron concentration in an intrinsic semiconductor is given the symbol n_i If we consider that the electrons are located (in terms of energy, remember) at the conduction-band edge E_c and the holes are located at the valence-band edge E_v, we get a nice little picture (Figure 4.10).

The concentration of holes at the valence-band edge is equal to the concentration of electrons at the conduction-band edge. Therefore the Fermi level E_f must lie near the middle of the bandgap (Figure 4.11).

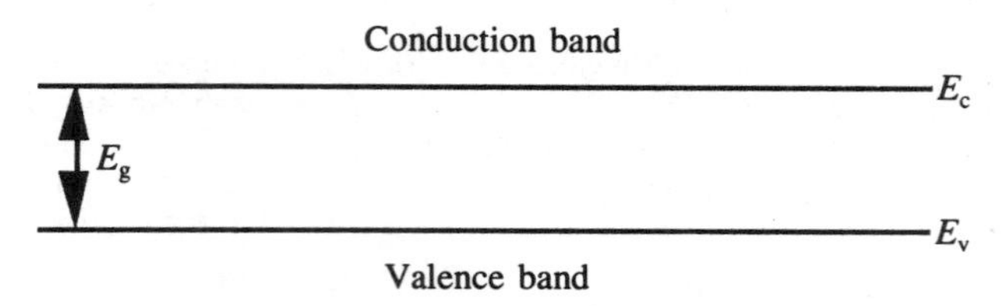

Figure 4.10 *Simple energy-band diagram for an intrinsic semiconductor.*

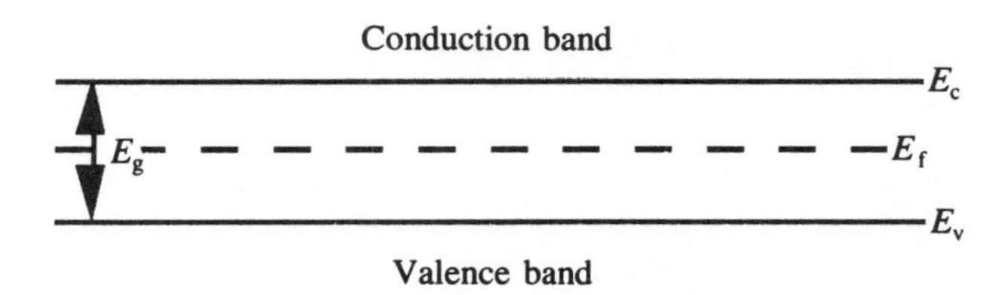

Figure 4.11 *Position of the Fermi level for an intrinsic semiconductor.*

It doesn't matter that the Fermi level is in the bandgap where electrons aren't allowed in an intrinsic material. The Fermi level just tells us that if there were a state at E_f, and *if* there were an electron to occupy it, then the probability of finding the electron in that state would be a half.

4.2.2 The Fermi level in extrinsic semiconductors

The Fermi level lies in different places in extrinsic semiconductors, depending on whether the material is n-type or p-type.

n-type

There is a high concentration of electrons in the conduction band compared to the hole concentration in the valence band, so the Fermi level must lie near the conduction band (Figure 4.12).

p-type

There is a high concentration of holes in the valence band compared with the electron concentration in the conduction band, so the Fermi level lies near the valence band (Figure 4.13).

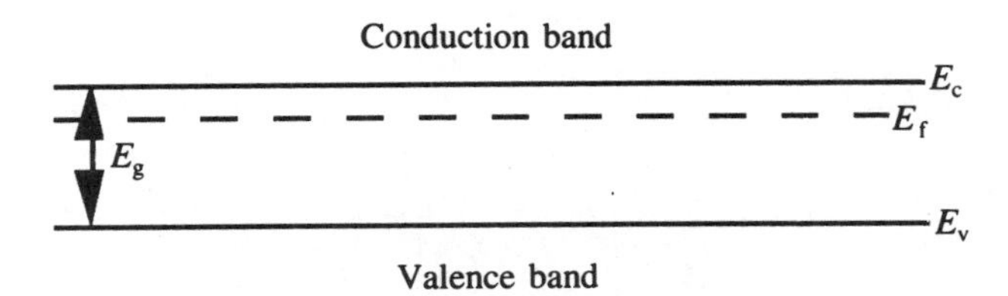

Figure 4.12 *Position of the Fermi level for an n-type semiconductor.*

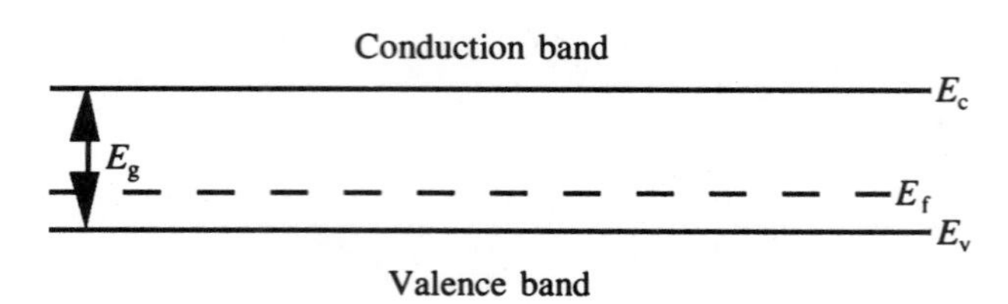

Figure 4.13 *Position of the Fermi level for a p-type semiconductor.*

4.3 Work function and electron affinity

4.3.1 Work function

A metal has a Fermi level E_{fm}; we've seen how the Fermi level of a semiconductor is the energy at which the probability of occupancy is 0.5 and the principle is the same in a metal. However, the energy-band diagram for a metal looks different to that of a semiconductor because, in the metal, all available states are filled up to the Fermi level (Figure 4.14).

The work function ϕ_m of a metal is the potential required to excite an electron from the Fermi level to the vacuum level (i.e. the so-called vacuum level outside the atom). It can be represented on an electron-energy scale in the same way that semiconductor energies can be shown on a band diagram. Figure 4.15 illustrates how an energy $e\phi_m$ is required to remove an electron at the Fermi level to the vacuum outside the metal, as demonstrated by the photoelectric effect (Section 1.5). Typical values of metal work function are 4.3 eV for aluminium and 4.8 eV for gold.

Figure 4.16 shows how this idea of work function applies to semiconductors. The work function in the semiconductor is the same as for the metal — it's the energy

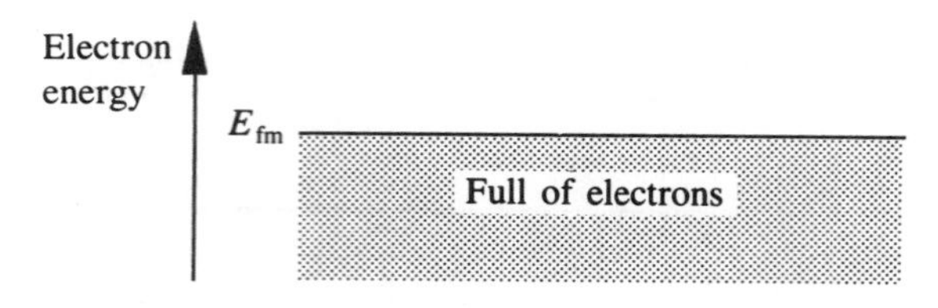

Figure 4.14 *Energy-band diagram for a metal, showing the Fermi level E_{fm}.*

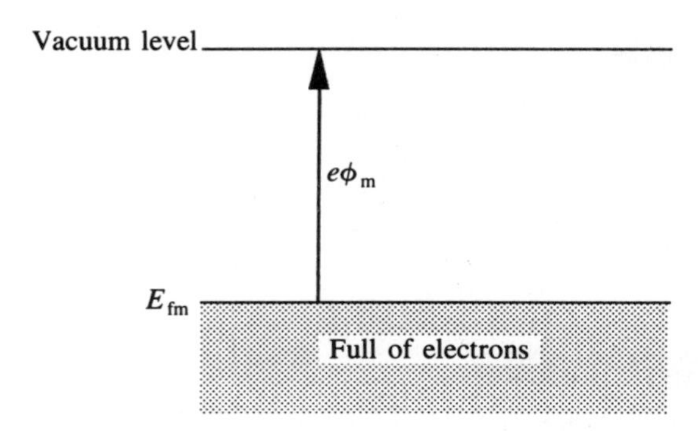

Figure 4.15 *Energy-band diagram for a metal, showing the work function $e\phi_m$.*

required to remove an electron from the semiconductor Fermi level E_{fs} to the vacuum outside the sample.

SELF-ASSESSMENT QUESTION 4.5

Note that the work function $e\phi$ in eV is equivalent to work function ϕ in volts. If ϕ is the potential required to excite an electron from the Fermi level E_f to the vacuum level, then $e\phi$ is the energy required to do so. Also note that e here is our usual e — the electronic charge, not the exponential function. What, then, is the voltage required to excite an electron from the

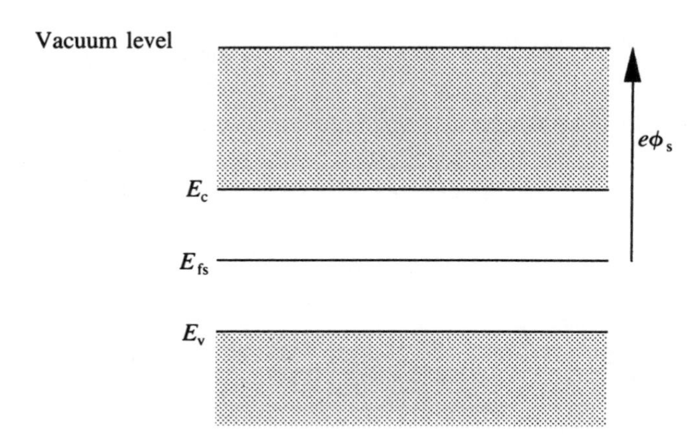

Figure 4.16 *Energy-band diagram for a semiconductor, showing the work function $e\phi_s$.*

Fermi level to the vacuum level in a sample of aluminium for which the work function is 4.3 eV?

4.3.2 *Electron affinity*

Another term used to describe semiconductors is the electron affinity χ. (This is the Greek letter chi and should not be confused with x). The electron affinity can be shown on the band diagram. It's the energy required to remove an electron from the bottom of the conduction band E_c to the vacuum level (Figure 4.17). The electron affinity $e\chi$ for silicon is approximately 1.4 eV. Other examples include 1.23 eV (germanium) and 1.26 eV (diamond).

SELF-ASSESSMENT QUESTION 4.6

What is the energy required to excite an electron from the conduction-band edge E_c to the vacuum level in a sample of silicon for which the electron affinity is 1.4 V?

4.4 Carrier transport in semiconductors

The way electrons and holes carry current is important. Engineers design devices to have particular electrical properties; this means that engineers have to control the currents that can flow in the devices. You've already seen that there are two types of charge carrier: the electron and the hole. Now we'll take a more detailed look at how these move around.

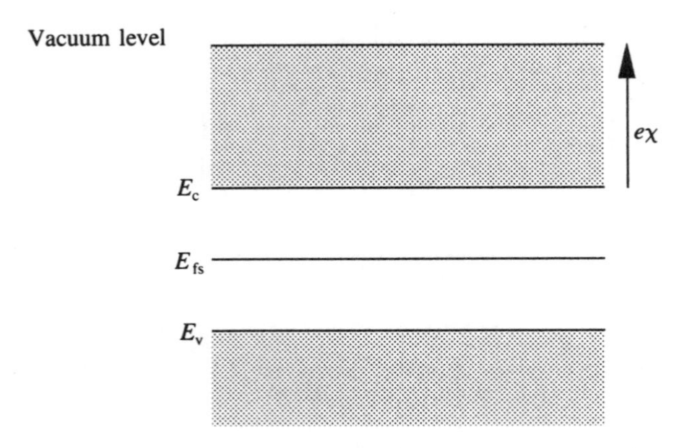

Figure 4.17 *Energy-band diagram for a semiconductor, showing electron affinity $e\chi$.*

4.4.1 **Motion of an electron in a crystalline solid**

Consider a sample of semiconductor material. It doesn't matter what sort it is, but suppose it's at a high enough temperature to contain free electrons. Those free electrons will be in constant motion and will wander around the lattice in a totally random manner — an electron is just as likely to be in one place as another (Figure 4.18). A current is just a flow of electrons, so whenever an electron is moving you may expect that a current will be formed. In our sample, however, the number of electrons is large and they are all travelling randomly in all directions. This means that the current caused by electrons travelling in a particular direction will be balanced by the electrons travelling in exactly the opposite direction. This applies throughout the sample, so there is no net current flow. The electrons move very fast during their random motion with a velocity depending on their temperature: the higher the temperature, the higher the thermal velocity v_{thermal}. We can write an expression for v_{thermal} (equation 4.1):

$$v_{\text{thermal}} = \left(\frac{3kT}{m_o}\right)^{1/2} \tag{4.1}$$

where k = Boltzmann's constant, T = temperature in kelvin K, m_o = rest mass of the electron = 9.11×10^{-31} kg.

$$k = \text{Boltzmann's constant} = 1.381 \times 10^{-23}\text{J K}^{-1}$$

SELF-ASSESSMENT QUESTION 4.7

According to equation (4.1), what would be the thermal velocity of an electron travelling in a material at room temperature?

Now suppose an electric field $\mathcal{E}$ is applied across the sample. How do the electrons move now? The basic random nature of the motion is retained, but the body of electrons moves towards the positive side of the sample (Figure 4.19). An electron is now more likely to be nearer the positive side than further away from it. This net movement of the body of electrons towards the positive pole (i.e. in a direction opposite to that of the electric field $\mathcal{E}$) is called drift. The drift velocity v_{drift} is less than the thermal velocity v_{thermal}. It's important to remember that the solid through which the electrons are travelling is crystalline. This means that electrons will collide with displaced or large atoms in the lattice; these collisions have to be considered when designing devices because they can slow down carrier transport. We can define a quantity τ called the mean free time, being the average time an electron moves between collisions. This enables us to write an expression for v_{drift} (equation 4.2)):

$$v_{\text{drift}} = \left(\frac{e\tau}{m_o}\right)\mathcal{E} \tag{4.2}$$

where e = charge on the electron, $\mathcal{E}$ = electric field, m_o = rest mass of the electron.

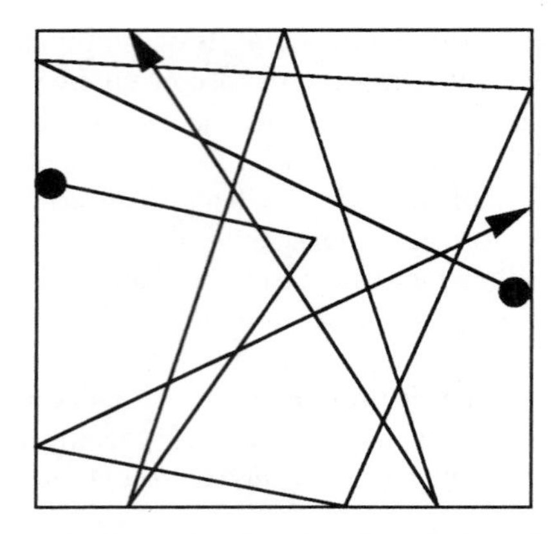

Figure 4.18 *Random thermal motion of two free electrons in a crystal lattice.*

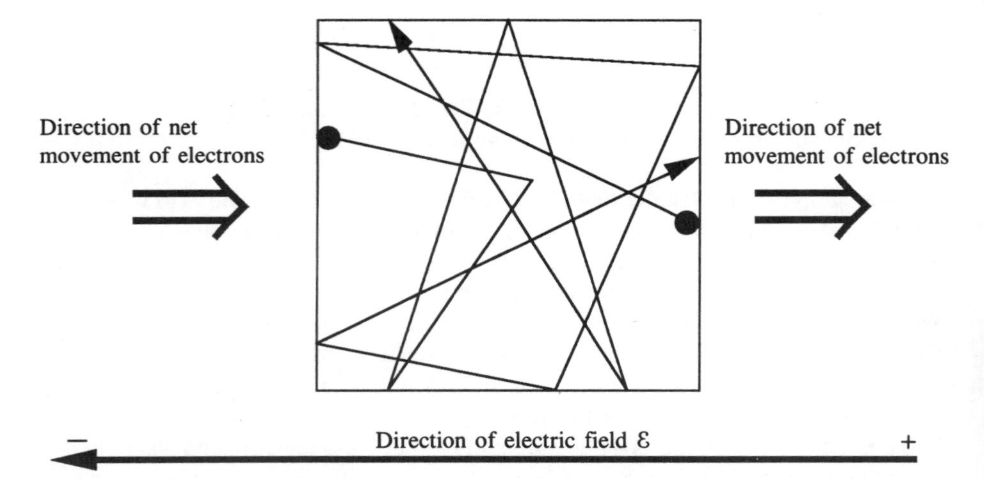

Figure 4.19 *Sketch illustrating the net movement of a body of electrons in a direction opposite to the applied electric field ℰ.*

4.4.2 Drift

Inspection of equation (4.2) enables us to define a new parameter μ, such that

$$v_{\mathrm{drift}} = \mu \mathcal{E} \tag{4.3}$$

where

$$\mu = \frac{e\tau}{m_o} \tag{4.4}$$

μ depends on the mean free time between collisions, τ. The parameter μ is very important because it indicates how easily electrons and holes can move through the lattice of the material. For this reason μ is called the mobility — the higher the mobility, the easier the movement. High mobility is often desired because it leads to higher conductivity, high current levels and faster response times.

SELF-ASSESSMENT QUESTION 4.8

According to equation (4.3), what drift velocity would an electron have in a material for which the mobility is $0.85\,\mathrm{m^2\,V^{-1}\,s^{-1}}$? Assume an electric field of $1\,\mathrm{V\,m^{-1}}$. Compare your value of v_{drift} with the value of v_{thermal} you determined for self-assessment question 4.7.

SELF-ASSESSMENT QUESTION 4.9

For the sample of self-assessment question 4.8, what would be the mean free time between collisions?

We could work out the current in the sample caused by the drift of the electrons, but a much more useful quantity is the current density J. This is because the amount of current flowing through a sample depends on the size of the sample, so it's hard to compare electrical properties between one sample and another when considering current. Current density, on the other hand, doesn't depend on the size of the sample so comparisons of J between two samples are meaningful. So consider the current density J. This is defined as the number of electrons crossing a unit area of material in unit time multiplied by the charge on the electron. Its units are therefore $\mathrm{A\,m^{-2}}$. If n is the concentration of electrons per unit volume,

$$J_{\mathrm{drift}} = nev_{\mathrm{drift}} \tag{4.5}$$

Hence J turns out to be proportional to the applied electric field $\mathcal{E}$:

$$J_{\mathrm{drift}} = ne\mu\mathcal{E} \tag{4.6}$$

$$J_{\mathrm{drift}} = \sigma\mathcal{E} \tag{4.7}$$

where σ is the constant of proportionality, called the conductivity. The units of conductivity are $\Omega^{-1}\,\mathrm{m^{-1}}$. Holes and electrons for a particular sample of material will have different values of mobility and conductivity: μ_{n}, μ_{p}, σ_{n}, σ_{p}. The subscript n denotes electrons, the subscript p denotes holes. Generally speaking, holes have lower mobilities than electrons. For this reason, high-speed devices use n-type rather than p-type materials to give high conductivities.

SELF-ASSESSMENT QUESTION 4.10

An electric field of $5\,\text{V cm}^{-1}$ is applied across a sample of semiconductor and the resulting current density is found to be $25\,\text{mA cm}^{-2}$. What is the conductivity of the sample?

Conductivity can be written:

$$\sigma = ne\mu_n \tag{4.8}$$

where e is the electronic charge (C), n is the concentration of electrons (m^{-3}), and μ_n is the electron drift mobility ($\text{m}^2\,\text{V}^{-1}\,\text{s}^{-1}$). Note the units of mobility, they're quite complex. Here mobility is expressed in $\text{m}^2\,\text{V}^{-1}\,\text{s}^{-1}$, but later on you'll see mobility expressed as $\text{cm}^2\,\text{V}^{-1}\,\text{s}^{-1}$.

SELF-ASSESSMENT QUESTION 4.11

An n-type semiconductor sample has an electron concentration of $2 \times 10^{22}\,\text{m}^{-3}$. If the conductivity is $2500\,\Omega^{-1}\,\text{m}^{-1}$, what is the electron drift mobility?

SELF-ASSESSMENT QUESTION 4.12

In self-assessment question 4.8, the sample has an electron drift mobility of $0.78\,\text{m}^2\,\text{V}^{-1}\,\text{s}^{-1}$. What is the electron drift mobility expressed in $\text{cm}^2\,\text{V}^{-1}\,\text{s}^{-1}$?

If the semiconductor sample contains both holes and electrons, the expression for conductivity will have two components representing the contribution to the conductivity from each type of carrier:

$$\sigma = en\mu_n + ep\mu_p \tag{4.9}$$

where p is the hole concentration (m^{-3}) and μ_p is the hole drift mobility ($\text{m}^2\,\text{V}^{-1}\,\text{s}^{-1}$). Examples of electron and hole mobilities are shown in Table 4.1 for various semiconductors at 300 K.

Table 4.1 *Electron and hole mobilities for common semiconductors*

	Ge	Si	GaAs
$\mu_n (\text{cm}^2\text{V}^{-1}\text{s}^{-1})$	3900	1350	8500
$\mu_p (\text{cm}^2\text{V}^{-1}\text{s}^{-1})$	1900	480	450

Resistivity ρ is often used instead of conductivity σ:

$$\rho = \frac{1}{\sigma}\ \Omega\,\text{m} \tag{4.10}$$

SELF-ASSESSMENT QUESTION 4.13

What is the conductivity of a sample of silicon which has a resistivity of 100 Ω cm?

The relationship between resistivity and resistance

Resistivity ρ is commonly used instead of resistance R by semiconductor-device engineers. This is because the resistance doesn't contain any information about the size of the sample: the size of the semiconductor sample can have a big effect on the size of the currents and voltages measured, so it's important to use parameters, such as resistivity, which aren't affected by size. Consider a rectangular slab of semiconductor shown in Figure 4.20.

The sample has two electrodes connected to it so that a voltage can be applied across it. (This applied voltage is $V_{applied}$.) This applied voltage will cause a current I to flow through the sample, from the positive electrode to the negative electrode. The resistance R of the sample is given by Ohm's law:

$$R = V_{applied}/I \ \Omega \tag{4.11}$$

The resistivity, however, is similar to the resistance but it takes the sample dimensions into account:

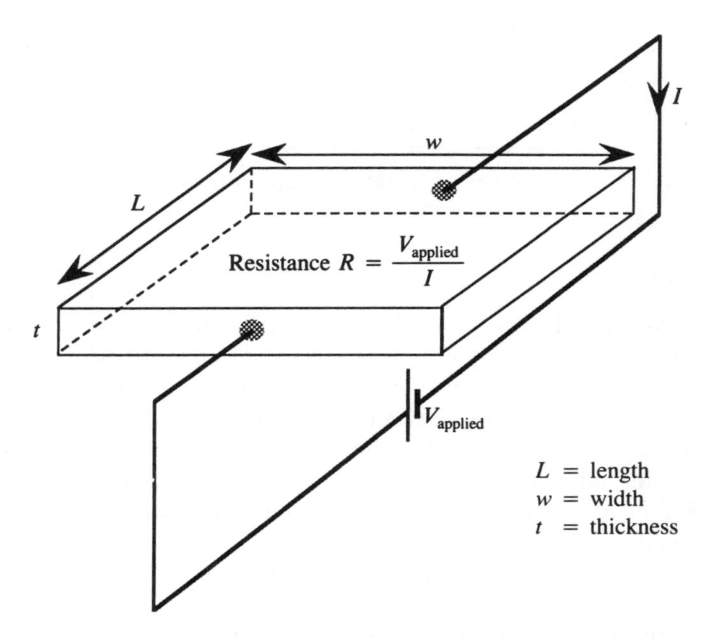

Figure 4.20 *A semiconductor sample set up to measure resistance.*

$$\rho = \frac{Rwt}{L} \quad \Omega\,\mathrm{m} \tag{4.12}$$

RESEARCH 4.1

Find the definition of resistivity and write it in your notebook.

SELF-ASSESSMENT QUESTION 4.14

A 12 V voltage is applied across a sample of semiconductor and a 60 mA current is measured through it. If the sample is 5 mm wide, 1 cm long, and 500 μm thick, what is its resistivity? Give your answer in Ω m and in Ω cm.

Table 4.2 lists some resistivities of some pure elemental conductors at room temperature.

Table 4.2 *Resistivities of some elemental conductors at room temperature*

Element		$\rho(\times 10^{-8}\,\Omega\,\mathrm{m})$
Ag	silver	1.59
Cu	copper	1.68
Au	gold	2.21
Al	aluminium	2.65
Na	sodium	4.77
W	tungsten	5.28
Pb	lead	20.8
Zr	zirconium	42.1

(*Source:* Lide, D.R. (ed.), *CRC Handbook of Chemistry and Physics*, 75th edn, CRC Press, 1994.)

4.4.3 **Diffusion**

So far we've assumed that the physical properties of a semiconductor sample are the same throughout its extent. Now we need to consider specimens in which the impurity doping is different in different regions of the specimens. Any such spatial variation or gradient in n or p leads to a net motion of the carriers from regions of high carrier concentration to regions of low carrier concentration. Charge carriers therefore will diffuse under the influence of a carrier-concentration gradient and this diffusion process will form a diffusion current. This represents an important charge-transport process in semiconductors.

Carriers in a semiconductor diffuse in a carrier-concentration gradient by random thermal motion and by scattering off the atoms making up the semiconductor lattice. We can calculate the rate at which the electrons diffuse in one dimension; this gives the electron (or hole) flux density $\Phi_n(x)$, which is the rate of electron flow in the $+x$ direction per unit area.

$$\Phi_n(x) = \frac{-D_n \, dn(x)}{dx} \tag{4.13}$$

This is known as the one-dimensional diffusion equation for electrons D_n is the electron diffusion coefficient ($m^2 s^{-1}$). The minus sign is included in the equation to indicate that the net motion of electrons due to diffusion is in the direction of decreasing electron concentration. This is the result we expect, since net diffusion occurs from regions of high carrier concentration to regions of low carrier concentration. Similarly, holes in a hole-concentration gradient move with a diffusion coefficient D_p according to the equation

$$\Phi_p(x) = \frac{-D_p \, dp(x)}{dx} \tag{4.14}$$

Note that the electron gradient is $dn(x)/dx$ and the hole gradient is $dp(x)/dx$.

The current density J is the current crossing a unit area, so this must be equal to the carrier flux density $\Phi(x)$ multiplied by the carrier charge:

$$J_{n(\text{diff.})} = e\Phi_n(x) \tag{4.15}$$

$$= \frac{eD_n \, dn(x)}{dx} \tag{4.16}$$

and

$$J_{p(\text{diff.})} = \frac{-eD_p \, dp(x)}{dx} \tag{4.17}$$

Table 4.3 shows some typical values of diffusion coefficients for holes and electrons in various semiconductors.

SELF-ASSESSMENT QUESTION 4.15

A sample of gallium arsenide has an electron diffusion coefficient of 220 $cm^2 s^{-1}$ and an electron concentration gradient of 1×10^4 cm^{-4} across its width, which is 1 cm. What electron diffusion current density would you expect from this sample?

Table 4.3 *Electron and hole diffusion coefficients for common semiconductors*

	$D_n(cm^2 s^{-1})$	$D_p(cm^2 s^{-1})$
Germanium Ge	100	50
Silicon Si	35	12.5
Gallium arsenide GaAs	220	10

4.4.4 ***The Einstein relationship***

There's a very useful equation which relates the two carrier-transport processes, drift and diffusion. It's called the Einstein relationship:

$$D = \frac{kT\mu}{e} \tag{4.18}$$

The existence of such a relationship isn't surprising because both diffusion and drift are statistical phenomena intimately linked with the same random processes of carrier motion. Note that the Einstein relationship contains both D, a diffusion parameter, and μ, a drift parameter. Boltzmann's constant k and the absolute temperature T relate to the random motion of the carriers.

SELF-ASSESSMENT QUESTION 4.16

If the electron drift mobility of a sample of semiconductor is 2500 $cm^2\,V^{-1}\,s^{-1}$ at room temperature, what would you expect the electron diffusion coefficient to be? Assume room temperature is 300 K.

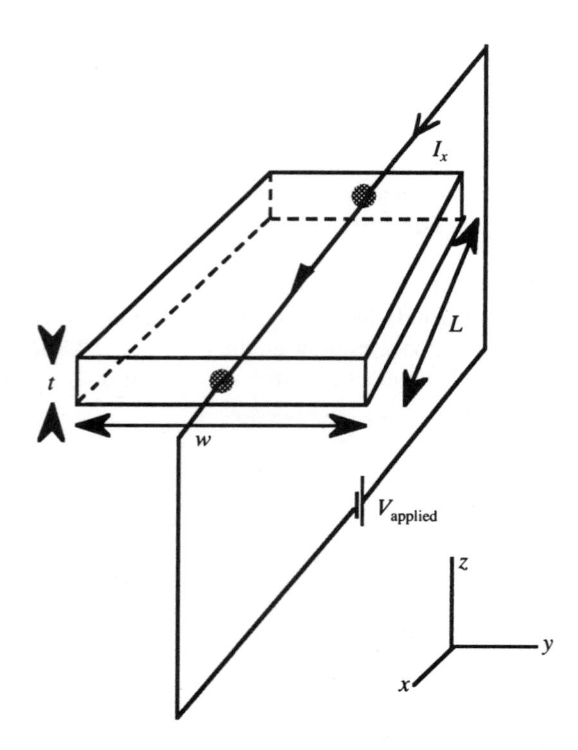

Figure 4.21 *A battery connected across a semiconductor sample to generate a current I_x through the sample.*

4.4.5 **The Hall effect**

The Hall effect is a phenomenon which is exploited by engineers to measure mobility and majority-carrier concentration. It's also used to determine the type of material: n or p.

Consider a rectangular slab of p-type semiconductor of length L, width w and thickness t. The slab is positioned so that the faces correspond to a set of coordinate axes x, y, z. Suppose it has metal electrodes deposited on two opposing faces (Figure 4.21). If a battery or power supply is connected by wires between these two electrodes a current will flow between them, from the positive side of the battery, through the semiconductor, and so to the negative pole of the battery. Call the voltage from the battery $V_{applied}$ and the current through the sample I_x. For a p-type slab, the current consists mostly of holes, so let's consider hole current only. Holes will travel in the same direction as the current I_x.

Suppose the sample is then placed in a magnetic field of flux density B_z such that the current I_x is orthogonal to B_z (Figure 4.22). Each hole making up the current I_x will then

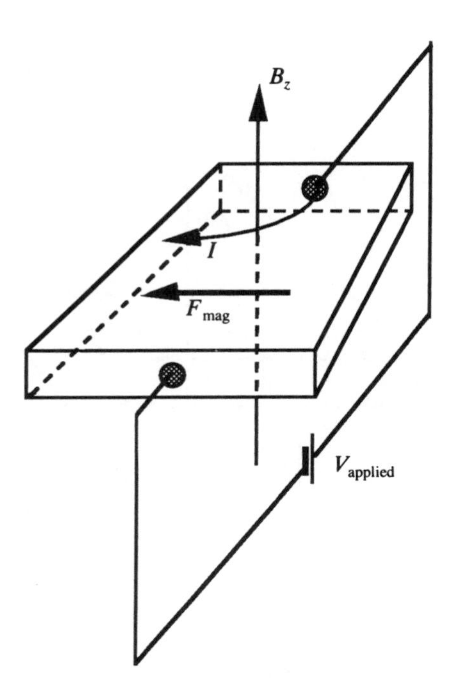

Figure 4.22 *The semiconductor sample placed in a magnetic field of flux density B_z, such that holes experience a force F_{mag}.*

experience a magnetic force F_{mag} acting in the $-y$ direction. This force will bend the hole current towards one of the side faces of the slab according to

$$F_{mag} = B_z e v_x \tag{4.19}$$

where e is the charge on the hole and v_x is the drift velocity of the holes.

This flow of holes onto the side of the slab causes an accumulation of holes at that face, hence the face becomes positively charged relative to its opposite face (Figure 4.23). The two sides of the slab now behave like the plates of a capacitor, and an electric field will be set up going from the positive face to the relatively negative face. This electric field is called the Hall field and is shown as $\mathcal{E}_H$ in Figure 4.23. The hole carriers in the sample will now experience an electric force F_{elec} in the $+y$ direction, according to

$$F_{elec} = e\mathcal{E}_H \tag{4.20}$$

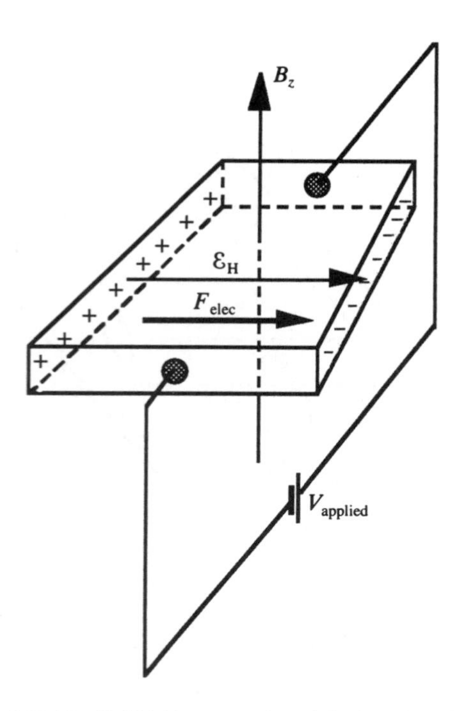

Figure 4.23 *The Hall field $\mathcal{E}_H$ set up in a slab of semiconductor.*

The Hall field increases in size until F_{elec} is equal in magnitude to F_{mag}. At this point the two forces balance each other exactly so that an individual hole does not experience a force and travels straight through the sample in the x-direction once more. Hence the original current I_x is restored (Figure 4.24). At this point the potential difference across the two faces in the y-direction is the Hall voltage V_H.

The Hall voltage V_H can be expressed in terms of the Hall field and the width of the sample:

$$V_H = \mathcal{E}_H w \tag{4.21}$$

This allows us to derive a useful expression which will allow the use of a technique to determine carrier type and carrier concentration in the slab. For an equation to be practically useful it should contain parameters which an engineer can measure. The measurable parameters in this case are: the slab dimensions L, w, t; the Hall voltage V_H; the current I_x; the magnetic flux density B_z.

At the balance point the two forces acting on each hole are equal in size and opposite in direction. This phenomenon is called the Hall effect.

$$F_{mag} = F_{elec} \tag{4.22}$$

Therefore,

$$Bzev_x = e\mathcal{E}_H \tag{4.23}$$

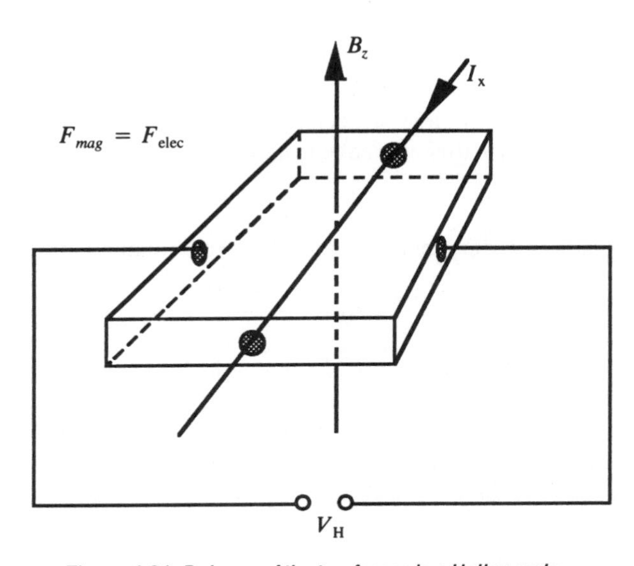

Figure 4.24 *Balance of the two forces in a Hall sample.*

Hence

$$\mathscr{E}_H = B_z v_x \tag{4.24}$$

We can write an expression for the hole drift velocity v_x based on equation (4.5):

$$v_x = \frac{J_x}{ep} \tag{4.25}$$

where J_x is the current density in the x-direction and p is the hole concentration. We can substitute equation (4.25) into equation (4.24) to give

$$\mathscr{E}_H = \frac{J_x B_z}{ep} \tag{4.26}$$

We can rearrange equation (4.26) to give an expression in terms of p:

$$p = \frac{J_x B_z}{e\mathscr{E}_H} \tag{4.27}$$

J_x is the current passing through the slab per unit area, therefore it can be written

$$J_x = \frac{I_x}{wt} \tag{4.28}$$

Equation (4.21) can be rearranged to give $\mathscr{E}_H$:

$$\mathscr{E}_H = V_H / w \tag{4.29}$$

Substituting equations (4.28) and (4.29) into equation (4.27) leads to

$$p = \frac{I_x B_z}{etV_H} \tag{4.30}$$

This is a very useful equation because I_x, B_z, t and V_H are all easy to measure. This means we can determine the hole concentration p_0 of any sample.

We can define a new parameter called the Hall coefficient R_H:

$$R_H = 1/ep \tag{4.31}$$

which enables us to rewrite equation (4.30) to give an expression for V_H:

$$V_H = \frac{I_x B_z R_H}{t} \tag{4.32}$$

The Hall coefficient R_H is useful because measuring it enables an engineer to know very quickly if a sample is n-type or p-type. If it is p-type, R_H will be positive and V_H will be positive, if the sample is n-type R_H will be negative and so will V_H. The polarity and size of V_H are determined just by reading a voltmeter.

We could determine the resistance R of the sample by measuring $V_{applied}$ and I_x:

$$R = V_{applied}/I_x \tag{4.33}$$

from Ohm's law. In terms of resistivity (equation (4.12)) this becomes

$$\rho = V_{\text{applied}} wt/I_x L \tag{4.34}$$

and, since conductivity σ is given by equation (4.10),

$$\sigma = 1/\rho = e\mu_p p \tag{4.35}$$

we can find an expression for the drift mobility μ_p:

$$\begin{aligned} \mu_p &= \sigma/ep \\ &= 1/\rho ep = R_H/\rho \end{aligned} \tag{4.36}$$

Similar results would be obtained for an n-type slab:

$$n = \frac{I_x B_z}{etV_H} \tag{4.37}$$

$$V_H = \frac{-I_x B_z}{etn} \tag{4.38}$$

$$\begin{aligned} \mu_n &= -\sigma/en \\ &= -1/\rho en \end{aligned} \tag{4.39}$$

The charge on an electron is negative ($-e$), so V_H and R_H are negative.

Hence the Hall effect is a useful phenomenon which can be exploited to enable the practical determination of carrier type, majority-carrier concentration and mobility. These parameters are all very important to device designers.

4.5 **Optical generation and photoconductivity**

Electron–hole pairs are generated when electrons in a semiconductor absorb energy. So far we've assumed that this carrier generation occurs because of the temperature of the sample, whereby energy in the form of heat is absorbed by the electrons. However, light energy will also be absorbed by the electrons to create electron–hole pairs. This process is called optical generation, and the resulting conductivity of the sample is called photoconductivity.

When a semiconductor of bandgap E_g is illuminated by light of wavelength λ, electron–hole pairs are generated providing the photons are absorbed by the material. For absorption to occur, the photon energy E_{ph} must be at least as large as the bandgap E_g (Figure 4.25).

$$E_{ph} \geq E_g$$

therefore

$$E_g \leq \frac{hc}{\lambda}$$

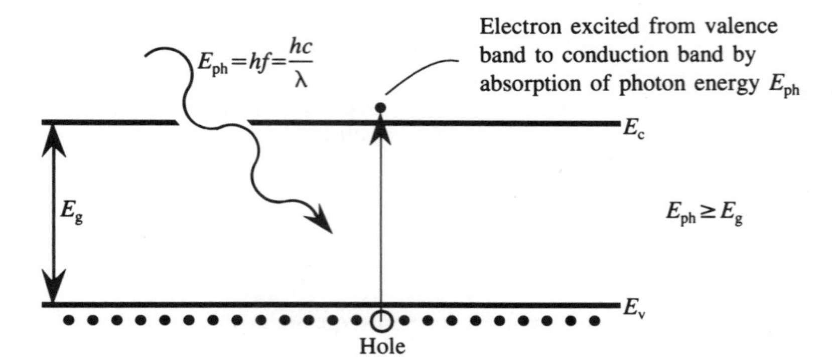

Figure 4.25 *Absorption of a photon of energy E_{ph} greater than or equal to the bandgap E_g to optically generate an electron–hole pair.*

for the optical generation of carriers. If $E_{ph} < E_g$ the photons pass straight through the semiconductor and are not absorbed (i.e. the material is transparent to these photons). The light does not need to be in the visible spectrum — infra-red and ultra-violet wavelengths are common in photoconductivity.

SELF-ASSESSMENT QUESTION 4.17

Would photons of wavelength 1300 nm, 644 nm and 480 nm be absorbed by gallium arsenide GaAs of bandgap 1.4 eV?

SELF-ASSESSMENT QUESTION 4.18

What 'colours' do the wavelengths listed in self-assessment question 4.17 correspond to?

At normal temperatures the optical generation of electron–hole pairs in a semiconductor occurs as well as thermal generation. This means that the optically generated carriers are 'extra' to those which are thermally generated. For this reason these carriers are called *excess* carriers. When excess electron–hole pairs are created the increase in the majority-carrier concentration is small compared to the increase in the minority-carrier concentration. Some optoelectronic devices, such as photodetectors, depend on the generation of excess carriers for their operation. You'll find out more about these in Chapter 7.

Once the excess carriers are generated by the absorption of photons these carriers will conduct in just the same way as the thermally generated carriers, and they will be free to wander around the material: when an electric field is applied across the semiconductor the holes in the valence band move towards the negative pole and the electrons in the conduction band move towards the positive pole.

4.6 Recombination

Once electron–hole pairs are generated they are free to take part in conduction. However, this freedom doesn't last for long. An electron which is excited into the conduction band will form a current under the influence of an applied electric field, but if the electron encounters a hole it will 'fall' into the hole, thereby returning to the valence band and removing itself from the possibility of conduction — until it is again excited from the valence band into the conduction band, when the process starts all over again. The falling into the hole is called recombination. The hole recombines also, of course, by having the electron fall into it. Figure 4.26(a) illustrates the recombination of an electron–hole pair across the bandgap. Recombination also occurs in extrinsic semiconductors between the conduction band and donor levels (Figure 4.26(b)), and between the valence band and acceptor levels (Figure 4.26(c)). Charge carriers in semiconductors are continually being generated and then gobbled up again by these electronic processes.

4.6.1 *Lifetime and diffusion length*

The time interval during which an individual hole or electron is free is very short, usually less than a microsecond. This time is called the lifetime and is usually denoted by the Greek letter tau τ. It varies from sample to sample, even of the same semiconductor. The lifetime is the average time that a carrier remains free between generation and recombination.

A quantity closely associated with lifetime is the diffusion length L. This is the average distance travelled by a free carrier during its lifetime. As with lifetime, diffusion length differs for holes and electrons, and from sample to sample.

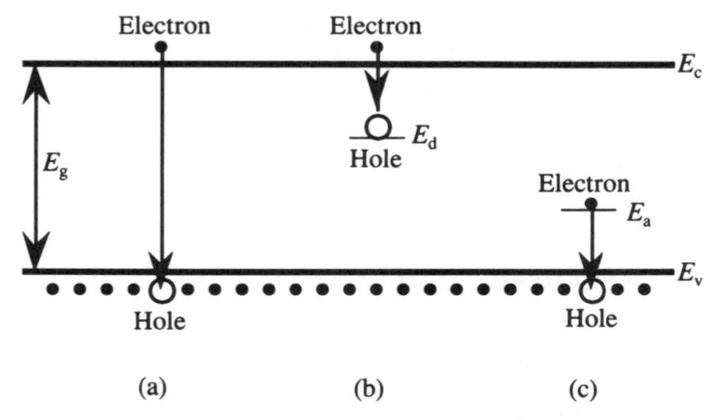

Figure 4.26 *Recombination in a semiconductor: (a) across the bandgap E_g, (b) between the conduction band and a donor level E_d, (c) between the valence band and an acceptor level E_a.*

Lifetime and diffusion length are related by the diffusion coefficient D as shown in equations (4.40) and (4.41).

For electrons,

$$L_n = (D_n \tau_n)^{1/2} \tag{4.40}$$

where L_n is the electron diffusion length, D_n is the electron diffusion coefficient and τ_n is the electron lifetime.

For holes,

$$L_p = (D_p \tau_p)^{1/2} \tag{4.41}$$

where L_p is the hole diffusion length, D_p is the hole diffusion coefficient and τ_p is the hole lifetime. If the units of D and τ are $m^2 s^{-1}$ and s respectively, then L will be in metres m.

4.7 Summary

Semiconductors can be divided into two groups: intrinsic and extrinsic. Intrinsic semiconductors are ideally pure, whereas extrinsic materials are purposely doped with donor atoms to become n-type or acceptor atoms to become p-type (Table 4.4). The semiconductors are doped to give larger possible currents and to enable control of those currents.

There are two types of current that can occur in a semiconductor: n-type and p-type. n-type current is made up of electrons, which are in the conduction band; p-type current is made up of holes, which are in the valence band. These dopant atoms are called donor atoms if they provide extra electrons (normally using pentavalent dopant atoms); they're called acceptor atoms if they provide extra holes (normally using trivalent dopant atoms). The sites (or energy levels, or energy states) these extra electrons and holes occupy are therefore called donor sites and acceptor sites. These sites are located in the bandgap, and very little energy is required to activate (excite) the carriers into or out of these sites.

Electrons move randomly through the semiconductor, colliding with the lattice. The thermal velocity of these electrons is high and depends on the temperature — the higher the temperature, the higher the thermal velocity.

Table 4.4 *Summary of extrinsic semiconductors*

	n-type	p-type
Majority carriers	Electrons in conduction band	Holes in valence band
Minority carriers	Holes in valence band	Electrons in conduction band
Dopant type	Donor (pentavalent)	Acceptor (trivalent)
Dopant level	Just below conduction band	Just above valence band

Under the influence of an electric field the population of carriers experiences a net movement in one direction called drift. Drift velocities are much lower than thermal velocities. The parameters mobility, resistivity and conductivity all indicate how good the carriers in the material are at moving around the lattice. High mobility and conductivity are generally desirable.

Carriers diffuse under the influence of a carrier-concentration gradient and a measurable parameter called the diffusion coefficient is used to indicate how much diffusion occurs in a semiconductor sample.

The total current in a semiconductor sample is the sum of electron drift current, electron diffusion current, hole drift current and hole diffusion current.

Equations for drift current density and diffusion current density have been derived. The Einstein relationship between drift mobility and diffusion coefficient is stated to show that drift and diffusion can be related.

The Hall effect is a phenomenon whereby a magnetic force can be introduced across a current-carrying sample of semiconductor by placing the sample in a magnetic field. The magnetic force leads to the formation of an equal and opposite electrical force. At balance, equations can be derived to enable the practical determination of carrier type, majority-carrier concentration and drift mobility.

Electron–hole pairs can be optically generated when photons of energy greater than the bandgap energy are absorbed by a semiconductor. The resulting carriers are called excess carriers and they conduct in the same way as thermally generated carriers. The conductivity caused by the excess carriers is called photoconductivity.

When electron–hole pairs are generated they will eventually recombine and remove themselves from any conduction processes. The average time a carrier is free before recombining is the lifetime and the average distance travelled by a carrier between generation and recombination is the diffusion length.

4.8 Tutorial questions

4.1 Describe, with the aid of sketches, how a hole current can be obtained in an intrinsic semiconductor.

4.2 Sketch energy-band diagrams to differentiate between n-type and p-type semiconductor materials.

4.3 Briefly describe what is meant by activation energy.

4.4 Why are semiconductors doped?

4.5 A particular sample of silicon has an electron affinity of 1.5 eV. If the Fermi level is located 0.55 eV below the conduction-band edge E_c, what is the work function for this sample of silicon?

4.6 A p-type semiconductor has a hole concentration of 5×10^{16} m^{-3}. If the hole conductivity is 32 Ω^{-1} m^{1}, what is the hole drift mobility?

4.7 If the electron drift mobility of a semiconductor is 2500 cm^2 V^{-1} s^{-1} at room temperature (300 K), what would you expect the electron diffusion coefficient to be?

4.8 The sketch (Figure 4.27) shows a sample of n-type silicon connected up for Hall-Effect measurements. It is placed in a 0.4 T magnetic field and a 50 mA current passes along the sample. Assume the sample thickness is 250 μm and the majority-carrier concentration n is 1×10^{20} cm^{-3}. What Hall voltage would you measure?

4.9 Assuming all dopant atoms are ionized, determine the following for a germanium Ge sample doped with 5×10^{18} atoms of antimony (Sb) cm^{-3}:

(i) electron concentration n;
(ii) hole concentration p;
(iii) conductivity σ;
(iv) resistivity ρ;
(v) resistance R.

Assume room temperature, plus the following:

$$\mu_n = 3900 \text{ cm}^2 \text{ V}^{-1} \text{ s}^{-1}$$
$$\mu_p = 1900 \text{ cm}^2 \text{ V}^{-1} \text{ s}^{-1}$$
$$n_i = 2.5 \times 10^{13} \text{ cm}^{-3}$$
$$\text{sample width } w = 2 \text{ mm}$$
$$\text{sample length } L = 0.5 \text{ cm}$$
$$\text{sample thickness } t = 250 \ \mu\text{m}$$

State any other assumptions you make, and explain what you're doing as you proceed with the calculations. What conclusions would you draw about the nature of the sample?

4.10 Describe how the Hall effect is set up in a rectangular sample of semiconductor. Use sketches and equations to aid your explanation.

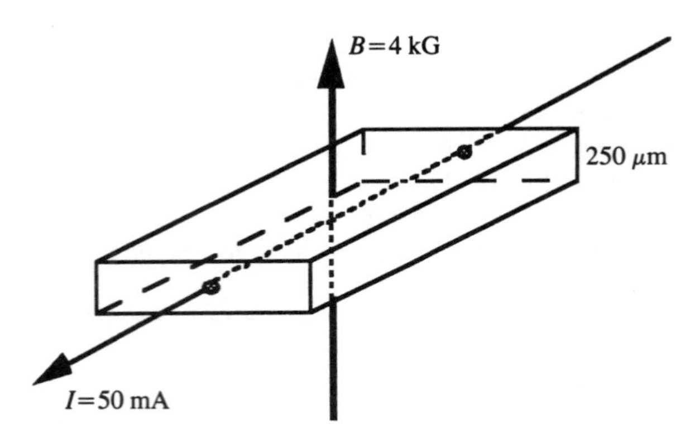

Figure 4.27 *Figure for tutorial question 4.8.*

4.11 A semiconductor sample is provided with four contacts and placed in a 10 kG magnetic field of flux density B_z. The current I_x is 5 mA. The sample dimensions are $w = 0.5$ cm, $t = 500$ μm, and $L = 1$ cm. The following data are measured: $V_H = 2.4$ mV and $V_{applied} = 2$ V. Is the sample n-type or p-type? Find the concentration of the majority carrier and its mobility.

4.12 A germanium sample is doped with 10^{19} antimony Sb atoms per cubic centimetre. What is the hole concentration p, the conductivity and the resistivity at 300 K? Assume the following:

$$\mu_n = 3900 \text{ cm}^2 \text{ V}^{-1} \text{ s}^{-1}$$

$$n_i = 2.5 \times 10^{13} \text{ cm}^{-3}$$

4.13 A sample of silicon is found to have a hole mobility of 480 $cm^2 V^{-1} s^{-1}$ at 300 K. What diffusion coefficient would you expect for this sample?

4.14 A semiconducting diamond sample is provided with four contacts and placed in a 7 kG magnetic field to enable Hall-effect measurements. A 4 V voltage is applied along the sample and the current through the sample is 6.4 μA. The sample dimensions are: width $w = 5$ mm, thickness $t = 100$ μm, and length L =1 cm. The Hall voltage is found to be 28 mV. Find the type and concentration of the majority carrier and its mobility. What do the results of your calculations tell you about this sample?

4.15 Find the resistivity and conductivity of intrinsic silicon. Assume a hole mobility of 480 $cm^2 V^{-1} s^{-1}$, an electron mobility of 1350 $cm^2 V^{-1} s^{-1}$, and an intrinsic carrier concentration n_i of 1.5×10^{10} cm^{-3}

4.16 What are the electron and hole concentrations for an n-type silicon sample doped with 10^{18} ions of arsenic (As) cm^{-3}? Assume an intrinsic carrier concentration in silicon of 1.5×10^{10} cm^{-3}. State any other assumptions you make.

4.9 **Suggested further reading**

Streetman B.G., *Solid State Electronic Devices*, 4th edn, Prentice Hall, 1995, pp. 81–92.

CHAPTER 5

The p–n junction

Aims and objectives

In this chapter the equations developed in Chapter 4 will be used to build a simple model of the p–n junction under zero-bias, forward-bias and reverse-bias conditions. Equations will be derived to show the relationships between contact potential, doping concentration, depletion width and capacitance under different bias conditions. Graphs showing the distributions of voltage, charge and electric field across the junction region will be sketched and equations written for the amount of charge in the junction region. The model will be extended to include asymmetrical p–n junctions where one side is more highly doped than the other, and to one-sided abrupt junctions where the doping on one side is insignificant. The chapter ends by briefly explaining a practical method of determining doping concentration and contact potential.

5.1 A simple model of the abrupt p–n junction

A p–n junction is formed in a *single* crystal of semiconductor by making one end of the crystal p-type by doping it with acceptor atoms and making the other end n-type by doping it with donor atoms. The junction is the region in the crystal where the p-type meets the n-type. The carrier-transport processes which take place across the junction lead to a switching action called rectification, such that the p–n junction passes a current in one direction but not in the opposite direction. This is the fundamental behaviour exhibited by all diodes.

The p–n junction is very important because it's the basis for many common electronic devices such as p–n diodes, optoelectronic devices such as light-emitting diodes and photodetectors, field-effect transistors and bipolar transistors. An understanding of what happens in the p–n junction is necessary to understand the operation of even the simplest electronic devices, such as rectifying diodes (you'll find out about their operation in Chapter 7). To do this we need a model which enables us to predict how a p–n junction will operate in a circuit.

5.1.1 Assumptions

We're going to model what happens at a p–n junction so that we can predict device performance. To do this it's important to state any assumptions at the outset. This is because they affect what comes afterwards. In this case, we're going to consider a simplified p–n junction to make it easier to understand. Even though the model we'll use is simplified it's a very good model because it leads to important results which match actual observations.

We need to make two assumptions before we look at the p–n junction:

1 The junction is abrupt (therefore called an abrupt or step junction). This means the concentration of electrons on the n-side of the sample is constant across that side of the sample. The concentration of holes on the p-side of the sample is likewise constant across that side of the sample (Figure 5.1).
2 We'll assume one-dimensional current flow in samples of uniform cross-sectional area, i.e. charge carriers move only in one direction. This is done to simplify the equations.

5.1.2 ***The p–n junction at equilibrium***

First of all we should define what we mean by *equilibrium*. The equilibrium state is achieved by leaving the p–n junction without any external connections to the electrical world outside it, i.e. we leave the p–n junction disconnected. We often call the equilibrium condition the *zero-bias* condition because of this.

p- and n-type materials in isolation

Consider samples of doped semiconductor material. They'll be either p-type or n-type. We can represent these samples as little boxes called p and n, and we can describe the electronic characteristics of the samples by drawing an energy-band diagram for each (Figure 5.2).

What happens when these samples are put into intimate electrical contact? (This means that the p- and n-types would both be part of the same single crystal.) Remind yourself of the assumptions we made earlier before you go any further.

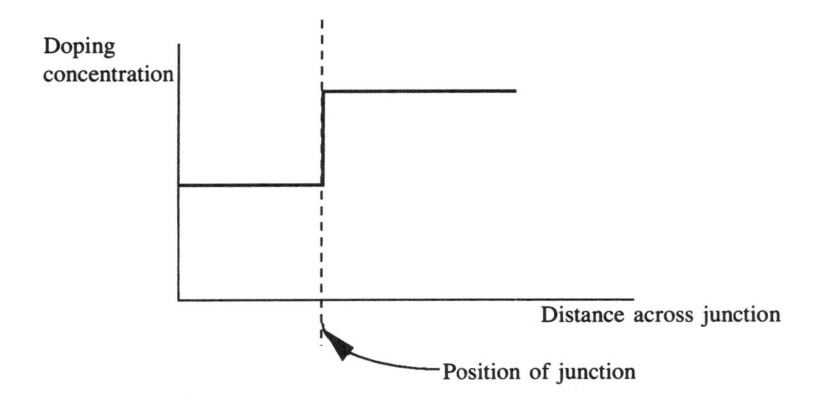

Figure 5.1 ***Illustration of an abrupt (step) junction.***

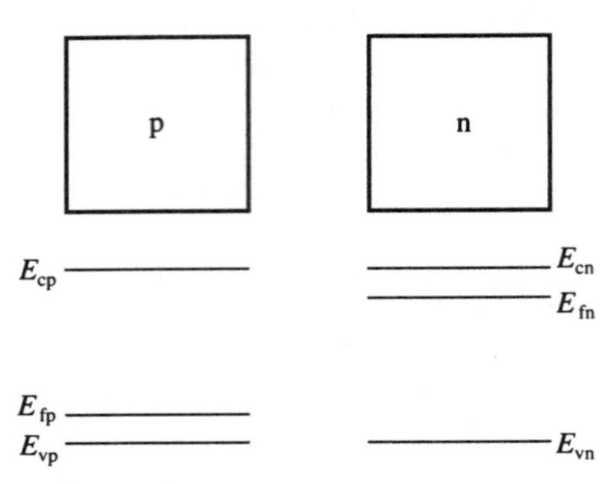

Figure 5.2 *p- and n-type materials in isolation.*

Modelling the abrupt junction at equilibrium

Note that, by equilibrium conditions, we mean that there is no external excitation (i.e. no voltage applied to the junction) and *no net current* flowing in the junction.

Consider the interface or junction between the p- and n-type materials (Figure 5.3). The p-type has a lot of holes and the n-type has a lot of electrons. Holes from the p-type diffuse across the junction into the n-type, and electrons from the n-type diffuse across the junction into the p-type. This is similar to the diffusion found in many systems, such as a partitioned box containing a different gas either side of the partition. When the partition is removed, both gases diffuse throughout the extent of the box, so that each gas is evenly distributed throughout the box (Figure 5.4).

Diffusion is always brought about by a concentration gradient, and in the case of the p–n junction it's the carrier-concentration gradient across the junction itself which causes the diffusion of carriers (Figure 5.5(a)). We get a hole diffusion current and an electron diffusion current across the junction (Figure 5.5(b)).

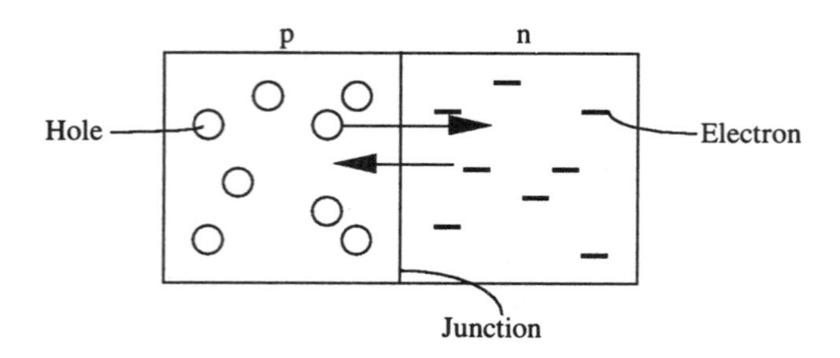

Figure 5.3 *Majority carriers in the p–n junction.*

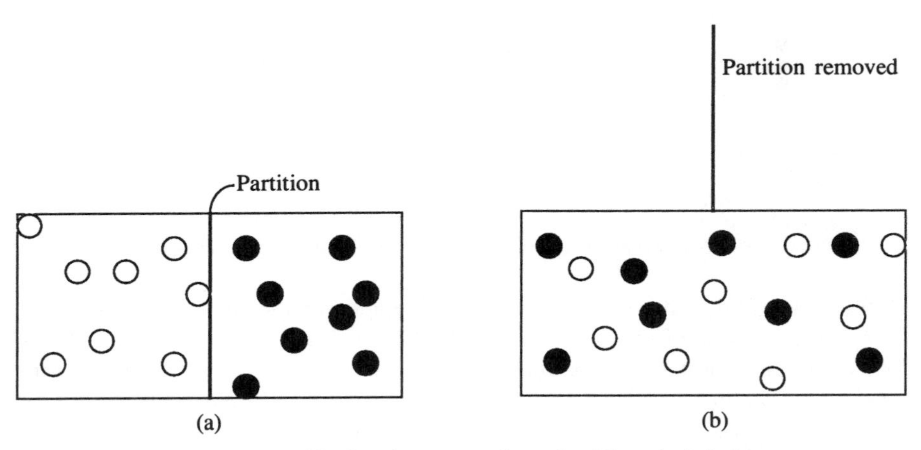

Figure 5.4 *The diffusion of two gases throughout the extent of a box.*

The example of the gases in the box is not a very good one. With the gases in the box, diffusion occurs *throughout* the box. In the case of a p–n junction, diffusion is brought to a halt because an electric field is built up which *opposes* the diffusion, so preventing the holes and electrons from diffusing through the whole sample. This is brought about as follows. As the diffusion proceeds across the junction, holes and electrons meet in the junction region and recombine. When an electron–hole pair is generated (Figure 5.6(a)) an electron is excited into the conduction band where it can conduct, and a hole is left in the valence band where it too can conduct. After a short time (called the *lifetime*) the electron in the conduction band will fall back into the hole it vacated in the valence band (Figure 5.6(b)). This falling back into a lower energy level or band to occupy a hole is recombination. The hole and electron no longer exist as charge carriers and so are removed from the conduction processes in the material. In the region of the junction where this recombination occurs, therefore, there are no charge carriers — this region is now depleted of charge carriers and is consequently called the depletion region. We'll call the width of this depletion region W, usually called the depletion width.

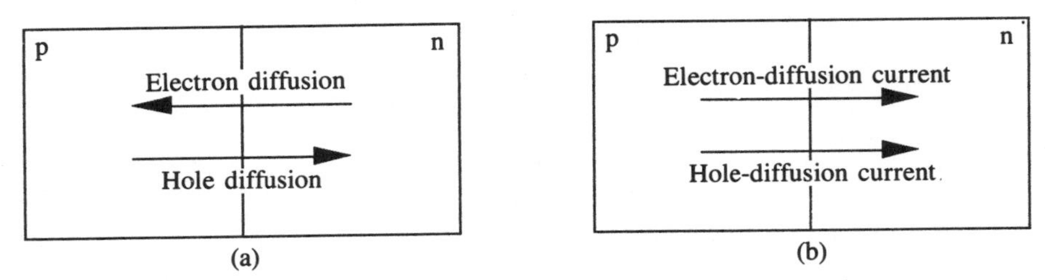

Figure 5.5 *Diffusion of carriers across the p–n junction and the resulting diffusion currents.*

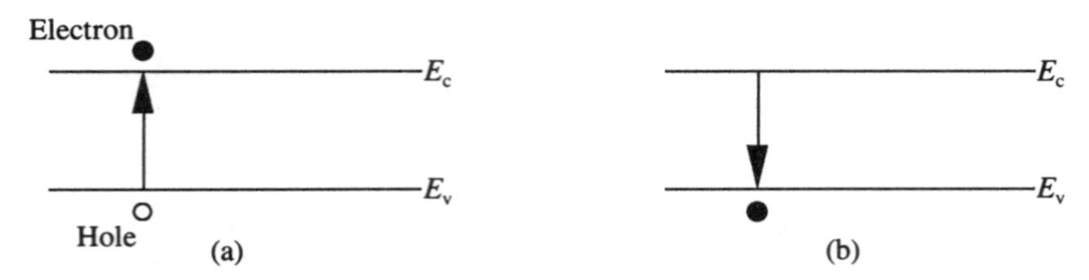

Figure 5.6 *Generation (a) and recombination (b) of an electron–hole pair.*

As a result of the diffusion the edges of the depletion region have layers of charge such that the p-side of the junction has a build-up of electrons associated with the ionized acceptor atoms, and the n-side has a build-up of holes associated with the ionized donor atoms. These layers are often referred to as space charge, and so the depletion region is sometimes also called the space-charge region. Hence one side of this region is negatively charged and the other is positively charged because of diffusion. Figure 5.7 shows a p–n junction with a depletion region in the centre of the sample, the edges of which show the space charge.

At each edge of the depletion region there's a build-up of charge, electrons on one side and holes on the other. This gives an electric field $\mathcal{E}$, called the built-in electric field. The direction of $\mathcal{E}$ is from the positive space-charge to the negative space-charge. Thus $\mathcal{E}$ is in the direction opposite to that of the diffusion current for each type of carrier (Figure 5.8). Hence the built-in electric field $\mathcal{E}$ creates a drift component from n to p, opposing the diffusion current. It is this opposition to the diffusion current that prevents diffusion occurring throughout the p–n junction sample. Carriers will then be transported across the junction by drift: i.e. electrons will drift from the p-type to the n-type, and holes will drift from the n-type to the p-type. Remember that holes and electrons drift under the influence of an applied electric field (Section 4.4). It's also important to remember at this point that doped material contains minority carriers.

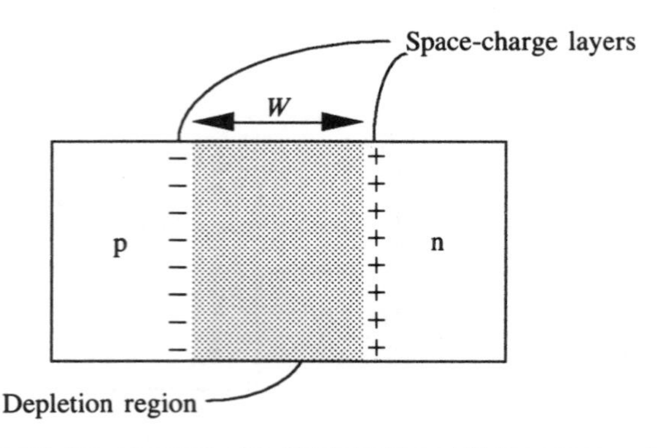

Figure 5.7 *A p–n junction, showing depletion width W and space charge.*

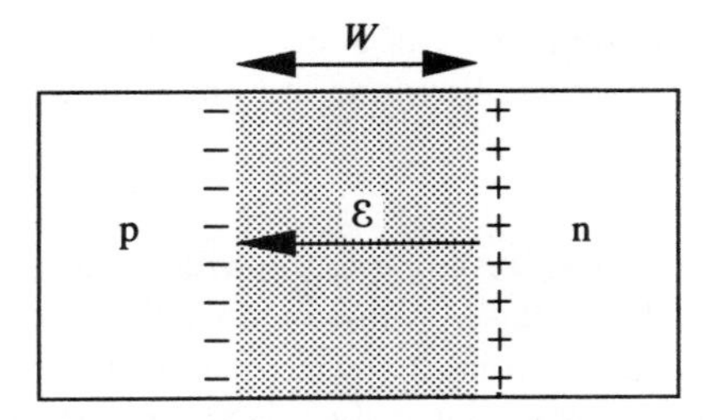

Figure 5.8 *A p–n junction, showing the direction of the built-in electric field ℰ across the depletion width W.*

Hence the majority carriers take part in diffusion, and the minority carriers take part in drift. Figure 5.9 illustrates the flow of carriers across the junction and also the flow of current across the junction — remember electrons move in the opposite direction to conventional current. However, because we know that there's no net current across the junction at equilibrium, the drift current must exactly cancel the diffusion current. Also, the drift and diffusion currents must cancel for each type of carrier. This means we can write

$$J_{p(drift)} + J_{p(diff.)} = 0 \tag{5.1}$$
$$J_{n(drift)} + J_{n(diff.)} = 0$$

indicating that the built-in electric field ℰ builds up in such a way that there is always

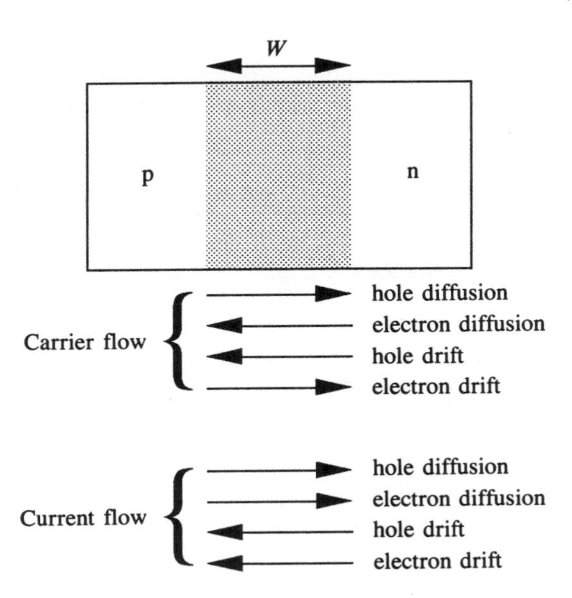

Figure 5.9 *The flow of carriers and currents across a p–n junction at equilibrium.*

zero net current at equilibrium. This is shown in Figure 5.9, where the arrows indicating the directions of carrier flow and current flow are all the same length.

Contact potential V_o

The electric field $\mathscr{E}$ appears across W, the depletion width. There is therefore a potential difference across W. We call this equilibrium potential difference the contact potential V_o. If we consider the potential from the space charge (i.e. electrostatic potential), there is a gradient between V_n (potential on the n-side caused by build-up of holes) and V_p (potential on p-side caused by build-up of electrons) (Figure 5.10). We can write

$$\mathscr{E}(x) = -\frac{dV(x)}{dx} \tag{5.2}$$

where V = electrostatic potential, x = distance through crystal.

Note that $\mathscr{E}(x)$ is written with reference to the x-direction, and this $\mathscr{E}$ is in the $-x$-direction (Figure 5.8). We assume that $\mathscr{E} = 0$ outside W, i.e. we assume that the regions outside the depletion region are electrically neutral. Note that

$$V_o = V_n - V_p \tag{5.3}$$

The contact potential V_o is a built-in potential barrier across the depletion width W. The existence of this potential barrier means that the p–n junction will act as a rectifier, i.e. it will act as a diode, allowing current to pass in one direction only. Note that the contact potential V_o cannot be measured directly. Any probes placed across the junction will form their own contact potentials which cancel out V_o.

RESEARCH 5.1

Look up some typical values of contact potential V_o in the available literature, and record them in your notebook. Note which materials they refer to.

Energy bands

Figure 5.11 shows the energy-band structure across the p–n junction. The valence and conduction bands are higher on the p-side of the junction than on the n-side by the amount eV_o. Note that the separation at equilibrium is just enough to make the Fermi

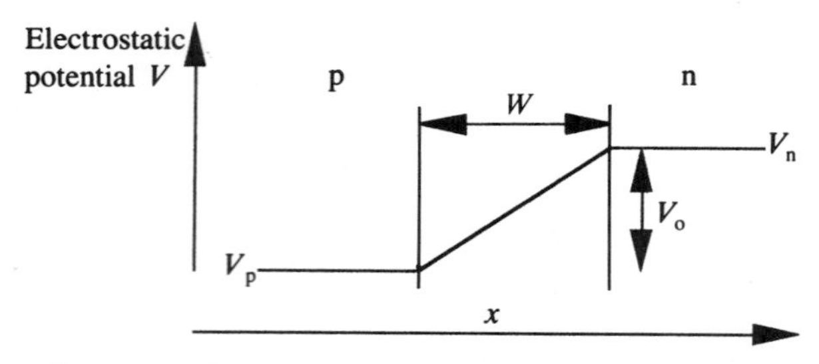

Figure 5.10 *Contact potential V_o across the depletion width.*

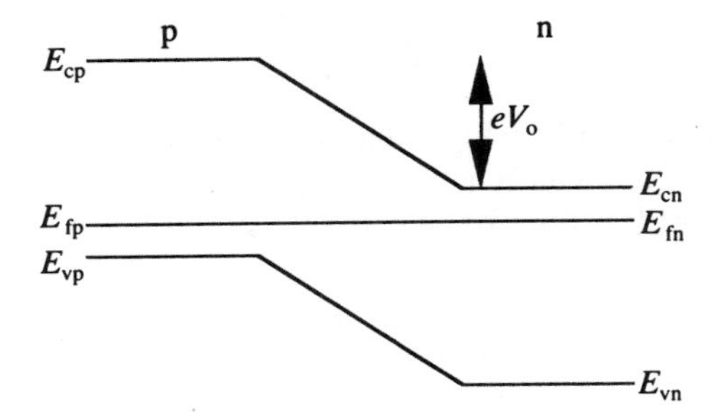

Figure 5.11 *Energy-band diagram for a p–n junction at equilibrium.*

level uniform throughout the region. See Appendix B for proof of this.

Relationship between V_o and doping concentration

It would be useful to be able to relate V_o to the doping concentration on each side of the junction, because engineers could then produce specific doping concentrations to give a desired barrier height and therefore desired device characteristics (such as diode threshold voltage). The height of the barrier V_o will determine the diode characteristic for any device made from a p–n junction. To find this relationship, we must consider equilibrium in terms of the equations describing drift and diffusion. Refer to Section 4.4 to refresh your memory about drift and diffusion.

In the p–n junction, the drift and diffusion components of the hole current just cancel at equilibrium, hence they give zero current density when added together:

$$e\Big(\underbrace{\mu_p p(x)\mathscr{E}(x)}_{\text{drift}} - \underbrace{D_p \frac{dp(x)}{dx}}_{\text{diffusion}}\Big) = 0 \tag{5.4}$$

Note the minus sign! This equation can be rearranged to give

$$\mu_p \mathscr{E}(x) = \frac{D_p}{p(x)} \frac{dp(x)}{dx} \tag{5.5}$$

The electric field $\mathscr{E}$ can be written in terms of the gradient in the potential (equation (5.2)). Using this equation we can substitute into equation (5.5) to get

$$\frac{-\mu_p}{D_p} \frac{dV(x)}{dx} = \frac{1}{p(x)} \frac{dp(x)}{dx} \tag{5.6}$$

However, the Einstein relationship states (equation (4.18)):

$$D_p = \frac{kT\mu_p}{e}$$

Therefore, substituting for μ_p/D_p gives

$$\frac{-e}{kT}\frac{\mathrm{d}V(x)}{\mathrm{d}x}=\frac{1}{p(x)}\frac{\mathrm{d}p(x)}{\mathrm{d}x} \tag{5.7}$$

We can solve this equation by integration over appropriate limits, in this case V_p and V_n (the potential either side of the junction). Since we have assumed a one-dimensional geometry, p and V are functions of x only. Integration gives

$$\frac{-e}{kT}\int_{V_p}^{V_n}\mathrm{d}V=\int_{p_p}^{p_n}\frac{1}{p}\,\mathrm{d}p \tag{5.8}$$

which gives

$$\begin{aligned}\frac{-e}{kT}(V_n-V_p)&=\ln p_n-\ln p_p\\&=\ln\left(\frac{p_n}{p_p}\right)\end{aligned} \tag{5.9}$$

Substituting for V_o from equation (5.3) and rearranging, we obtain

$$V_o=\frac{kT}{e}\ln\left(\frac{p_p}{p_n}\right) \tag{5.10}$$

where p_n = hole concentration on n-side of the junction, p_p = hole concentration on p-side of the junction.

So now we have a useful equation which relates the contact potential (or potential-barrier height) to the hole concentrations on either side of the junction. Another useful form of this equation is found by taking exponents:

$$\frac{p_p}{p_n}=\exp\left(\frac{eV_o}{kT}\right) \tag{5.11}$$

We can also extend the equation to include electrons, such that

$$\frac{n_n}{n_p}=\frac{p_p}{p_n}=\exp\left(\frac{eV_o}{kT}\right) \tag{5.12}$$

where n_n = electron concentration on n-side of the junction, n_p = electron concentration on p-side of the junction.

The derivation can be followed through for electrons in a similar way to that for holes (see the Tutorial Questions at the end of this chapter).

Carrier flow in a p–n junction at equilibrium

Figure 5.12 shows the location of carriers represented on an energy-band diagram for the p–n junction at equilibrium. On the p-side the minority carriers (electrons) are located in the conduction band; there aren't many of them. On the same side there are many majority carriers (holes) in the valence band. On the n-side there are many majority electrons in the conduction band and relatively few minority holes in the

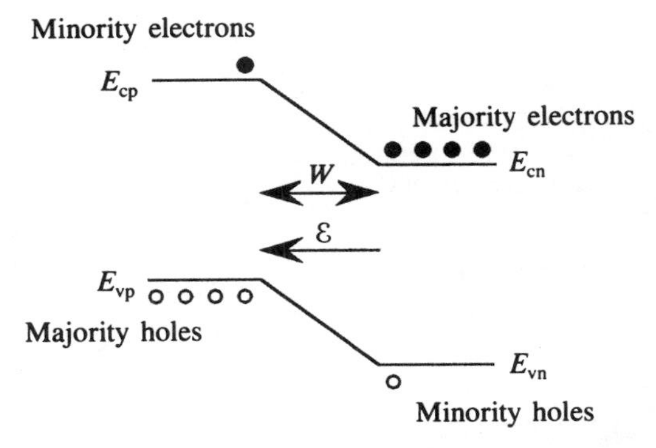

Figure 5.12 *Schematic diagram showing the location of carriers under zero bias (equilibrium conditions).*

valence band. There is a contact potential V_o volts, meaning that the energy barrier height is eV_o electron volts. Consider the p-side: the minority electron could easily fall down the energy slope across the depletion width because it would be energy-efficient to do so. This minority electron, in fact, would drift across the depletion width under the influence of the built-in electric field $\mathcal{E}$. Likewise, the minority hole on the n-side can easily drift across the depletion width under the influence of $\mathcal{E}$.

SELF-ASSESSMENT QUESTION 5.1

Why is it that the minority hole can easily cross the depletion width by drift, even though it appears to have to climb the barrier height eV_o?

Now consider the majority carriers represented in Figure 5.12. The majority electrons on the n-side have to gain eV_o electron volts of energy to climb the barrier, i.e. cross.the depletion width. Likewise the majority holes on the p-side also have to gain eV_o electron volts to cross the depletion width. These majority carriers can cross the depletion width only by diffusion, and the number crossing depends on the barrier height eV_o. Any carriers crossing the barrier constitute a current, hence there are four current components: hole diffusion, hole drift, electron diffusion and electron drift. Under zero bias (equilibrium conditions) these current components are all the same size, and the two drift-current components are from right to left in the figures and the two diffusion-current components are from left to right in the figures (Figures 5.9, 5.13). You might suppose that if the barrier height were reduced, the likelihood of more majority carriers crossing the depletion width would increase, and you'd be right.

SELF-ASSESSMENT QUESTION 5.2

The drift current is small, maybe about 1 nA. Why do you suppose it is so small?

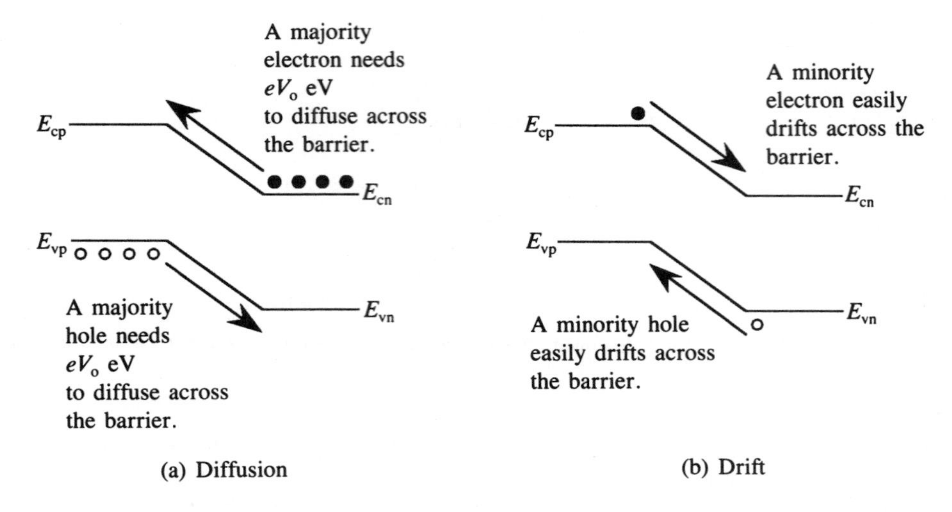

Figure 5.13 *Schematic diagram to illustrate diffusion and drift under zero bias.*

5.1.3 The p–n junction under forward bias

A p–n junction is forward biased when the p-side is made positive relative to the n-side (Figure 5.14). This results in several changes compared to the equilibrium condition:

- The depletion width is reduced.
- The built-in electric field decreases.
- The barrier height decreases.
- Diffusion current increases because the barrier height decreases, making it more likely that majority carriers will be able to cross the barrier.

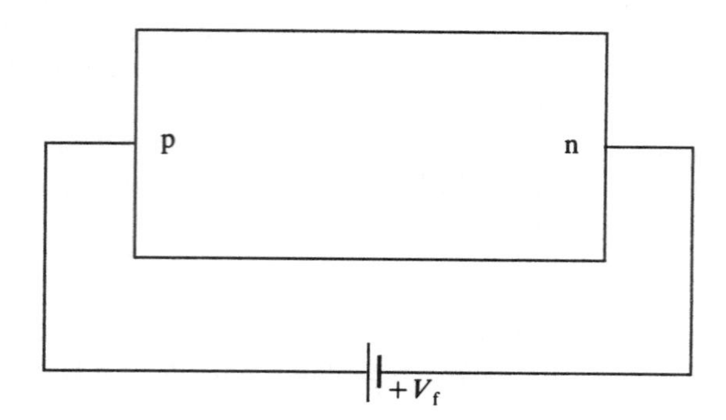

Figure 5.14 *A p–n junction connected in forward bias with bias voltage $+V_f$.*

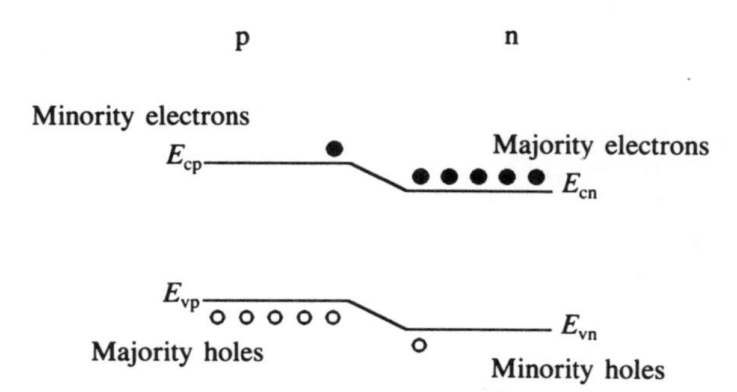

Figure 5.15 *Schematic diagram showing the location of carriers under forward bias.*

Note that drift current remains the same as in the equilibrium case (i.e. small). Figure 5.15 shows the location of the carriers on the energy-band diagram under forward-bias conditions.

SELF-ASSESSMENT QUESTION 5.3

Referring to Figure 5.15, can you explain why diffusion current increases and drift stays the same?

Figure 5.16 shows how much the barrier height is reduced under a forward-bias voltage $+V_f$. The bias voltage is effectively subtracted from the contact potential V_o, meaning that the voltage across the depletion width under forward bias is

$$V_o - (+V_f) = V_o - V_f \text{ volts}$$

This means that the barrier height is given by

$$e(V_o - V_f) \text{ eV}$$

If the bias voltage V_f was made equal to V_o then the barrier height would be zero. This is what happens when a diode is forward biased: a silicon diode, for example, needs a forward-bias voltage of about 0.7 V to get it to pass a large current. This is about the same size as the contact potential in a silicon p–n junction.

RESEARCH 5.2

Find values of threshold voltage V_T for real diodes made of various materials. List them all in your notebook and compare them with the values of contact potential V_o you found earlier (RESEARCH 5.1).

SELF-ASSESSMENT QUESTION 5.4

Would you expect V_o to be equal to the threshold voltage V_T?

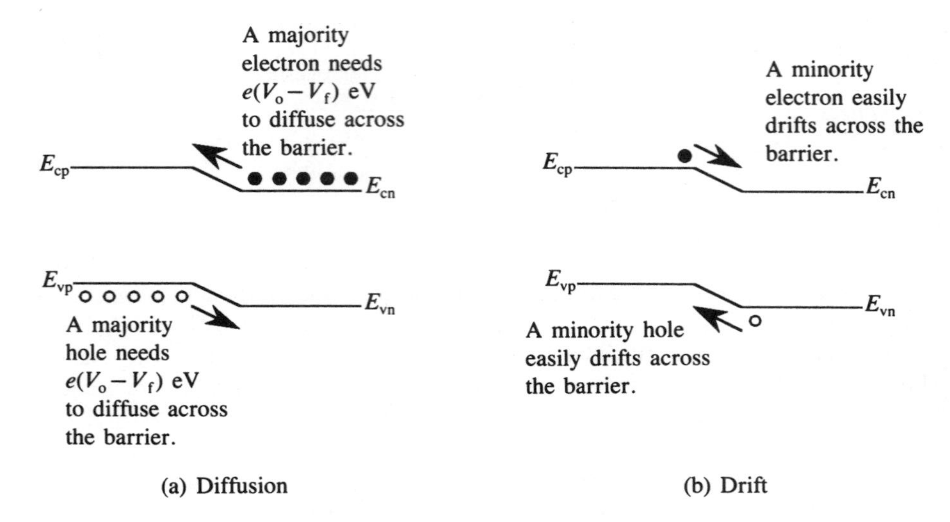

Figure 5.16 *Schematic diagram to illustrate diffusion and drift under forward bias.*

5.1.4 The p–n junction under reverse bias

A p–n junction is reverse biased when the p-side is made negative relative to the n-side (Figure 5.17). This results in several changes compared to the equilibrium condition:

- The depletion width is increased.
- The built-in electric field increases.
- The barrier height increases.
- Diffusion current decreases because the barrier height increases, making it more

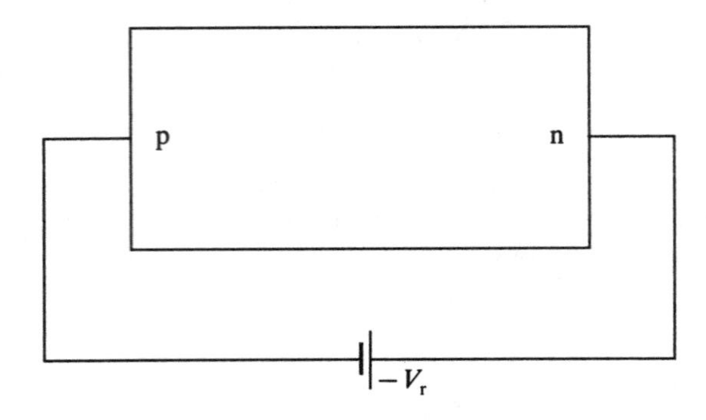

Figure 5.17 *A p–n junction connected in reverse bias with bias voltage $-V_r$.*

unlikely that majority carriers will be able to cross the barrier.

Note that drift current remains the same as in the equilibrium case (i.e. small). Figure 5.18 shows the location of the carriers on the energy-band diagram under reverse-bias conditions.

SELF-ASSESSMENT QUESTION 5.5

Referring to Figure 5.18, can you explain why diffusion current decreases and drift stays the same, compared to the equilibrium case?

Figure 5.19 shows how much the barrier height is increased under a reverse-bias voltage $-V_r$. The bias voltage is effectively subtracted from the contact potential V_o, meaning that the voltage across the depletion width under reverse bias is

$$V_o - (-V_r) = V_o + V_r \text{ volts}$$

This means that the barrier height is given by

$$e(V_o + V_r) \text{ eV}$$

It's important to realize that the increase in barrier height makes it very difficult for majority carriers to diffuse across the barrier, whereas the minority carriers can drift across easily. In reverse bias, therefore, the diffusion current is negligible and the drift current is very small.

RESEARCH 5.3

Find and sketch the *I–V* characteristic of a p–n diode.

SELF-ASSESSMENT QUESTION 5.6

Why does the diode conduct in only one direction?

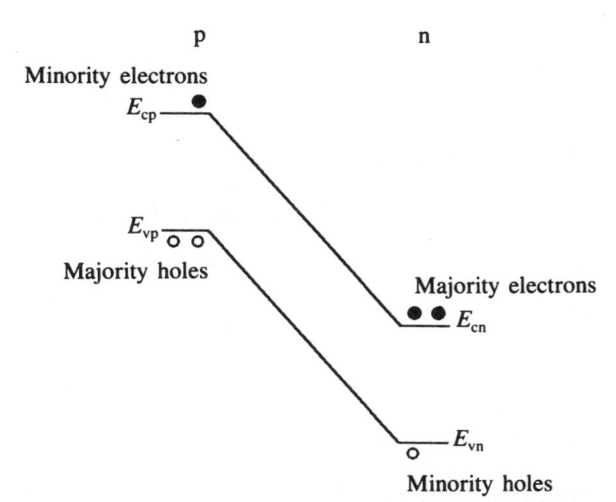

Figure 5.18 *Schematic diagram showing the location of carriers under reverse bias.*

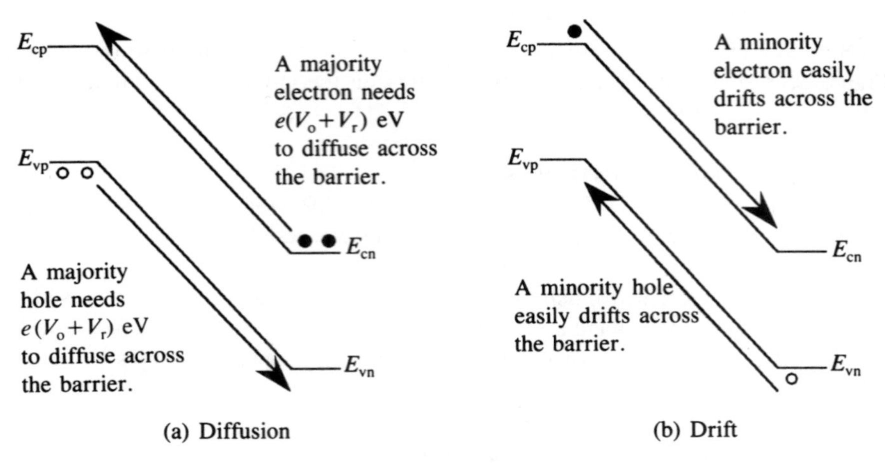

Figure 5.19 *Schematic diagram to illustrate diffusion and drift under reverse bias.*

RESEARCH 5.4

How large is the reverse current (often called reverse saturation current, or reverse leakage current or I_o) in a real p–n diode? Look up some values in the catalogues and handbooks, and write them in your notebook. Note what types of diode they are. How do these values compare to your observations of diode reverse current on a curve tracer[1] or oscilloscope? You may have thought that reverse current in a p–n diode was zero but oscilloscope scales usually aren't set to a sensitive enough scale to make a measurement possible.

SELF-ASSESSMENT QUESTION 5.7

Is the reverse current I_o caused by drift or by diffusion, and does it consist of majority or minority carriers?

5.2 p–n junction distributions

It's useful to be able to sketch graphs showing the distributions of the following across a p–n junction:

- voltage
- charge density
- electric field

1 A curve tracer is a piece of electrical equipment rather like a cathode-ray oscilloscope, but instead of giving a picture of wave amplitude versus time, it shows a plot of current versus voltage for both forward-bias voltages and reverse-bias voltages simultaneously. Curve tracers are therefore used to show the I–V characteristics of diodes (and transistors). Look in the available literature for a typical curve-tracer photograph of a p–n diode I–V characteristic. Alternatively use a curve tracer to inspect the characteristic.

There are very few charge carriers in the depletion region. Any carriers wandering into it are swept out by the built-in electric field $\mathcal{E}$. It's safe to assume therefore that all the charge in the depletion width W is provided by uncompensated donor and acceptor ions. We can plot this charge density as a function of x, where x is the distance across the junction from left to right (Figure 5.20). The figure also shows the notation I'm going to use to represent the penetration of the depletion region into the p-side and into the n-side. If $x = 0$ is the so-called metallurgical junction, then $-x_p$ is the distance the depletion region penetrates the p-side and x_n is the distance the depletion region penetrates the n-side, such that

$$W = |x_p| + |x_n| \tag{5.13}$$

Let Q_p be the quantity of charge between 0 and $-x_p$. The charge between 0 and x_n will be Q_n. Remember that the space charge on the n-side is positive because the donor ions are positive (i.e. they lose electrons) and that on the p-side the space charge is negative because the acceptor ions are negative (i.e. they gain electrons). Hence Q_n is positive and Q_p is negative. If we assume all the dopant atoms are ionized, Q_p will be the acceptor density N_a multiplied by the electronic charge $-e$, multiplied by the volume of the depletion region between 0 and $-x_p$:

$$Q_p = -eN_aAx_p$$

where A is the cross-sectional area of the junction (assumed to be constant throughout the sample). Likewise, on the n-side, the amount of charge between $x = 0$ and x_n will be Q_n, given by

$$Q_n = eN_dAx_n$$

Assuming space-charge neutrality we can then write

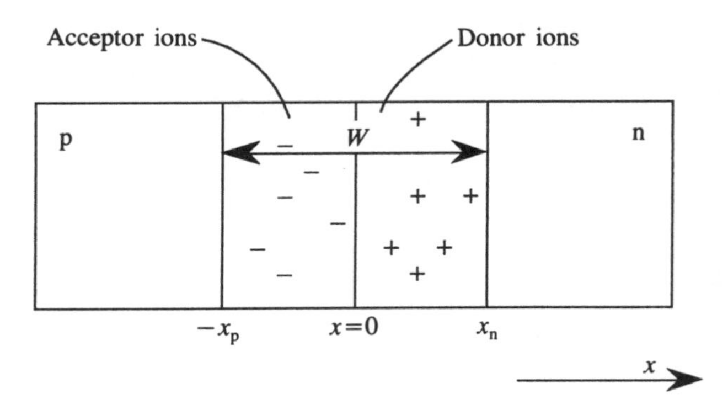

Figure 5.20 *Sketch of an abrupt symmetrical p–n junction, showing dopant ions in the depletion width W.*

$$|Q_n| = |Q_p|$$

or

$$eN_dAx_n = eN_aAx_p \tag{5.14}$$

We're assuming there are no charge carriers inside W and that the regions outside W are electrically neutral (this is called the *depletion approximation*).

The penetrations into the depletion region can be found using equation (5.14):

$$N_dx_n = N_ax_p$$

so

$$N_d/N_a = x_p/x_n$$

which leads to the following expressions for x_p and x_n:

$$x_p = N_dW/(N_a + N_d) \tag{5.15}$$

$$x_n = N_aW/(N_a + Nd) \tag{5.16}$$

Figure 5.21(a) is a plot of voltage distribution, where V_p is the potential at the edge of the depletion region on the p-side and V_n is the potential at the edge of the depletion region on the n-side. Note that V_p is less than V_n because V_p is relatively negative. Figure 5.21(b) is a plot of charge density (coulombs per unit volume) versus x, showing that the charges Q_n and Q_p are equal.

Figure 5.21(c) shows the electric-field distribution across the sample. This is obtained from Gauss' law which relates the amount of charge within a surface to the electric field. For the depletion region in a p–n junction this law can be written

$$\frac{d\mathcal{E}(x)}{dx} = \frac{e}{\epsilon_s}(N_d - N_a) \tag{5.17}$$

where $d\mathcal{E}(x)/dx$ is the electric-field distribution and ϵ_s is the semiconductor permittivity. We can write equation (5.17) for the p-side only and the n-side only, because on the p-side $N_d = 0$ and on the n-side $N_a = 0$.

On the p-side,

$$\frac{d\mathcal{E}(x)}{dx} = \frac{-eN_a}{\epsilon_s} \tag{5.18}$$

On the n-side,

$$\frac{d\mathcal{E}(x)}{dx} = \frac{eN_d}{\epsilon_s} \tag{5.19}$$

If you read about these equations (there are some suggestions at the end of this Chapter) you'll find that forms of these equations are named after Maxwell and

Poisson. Maxwell's equations and Poisson's equations are very important for a deep understanding of semiconductor physics but are beyond the scope of this book. Note that the semiconductor permittivity $\mathcal{E}_s$ is usually expressed in the form of equation (5.20), where $\mathcal{E}_r$ is the relative permittivity of the semiconductor (a ratio) and $\mathcal{E}_o$ is the permittivity of free space.

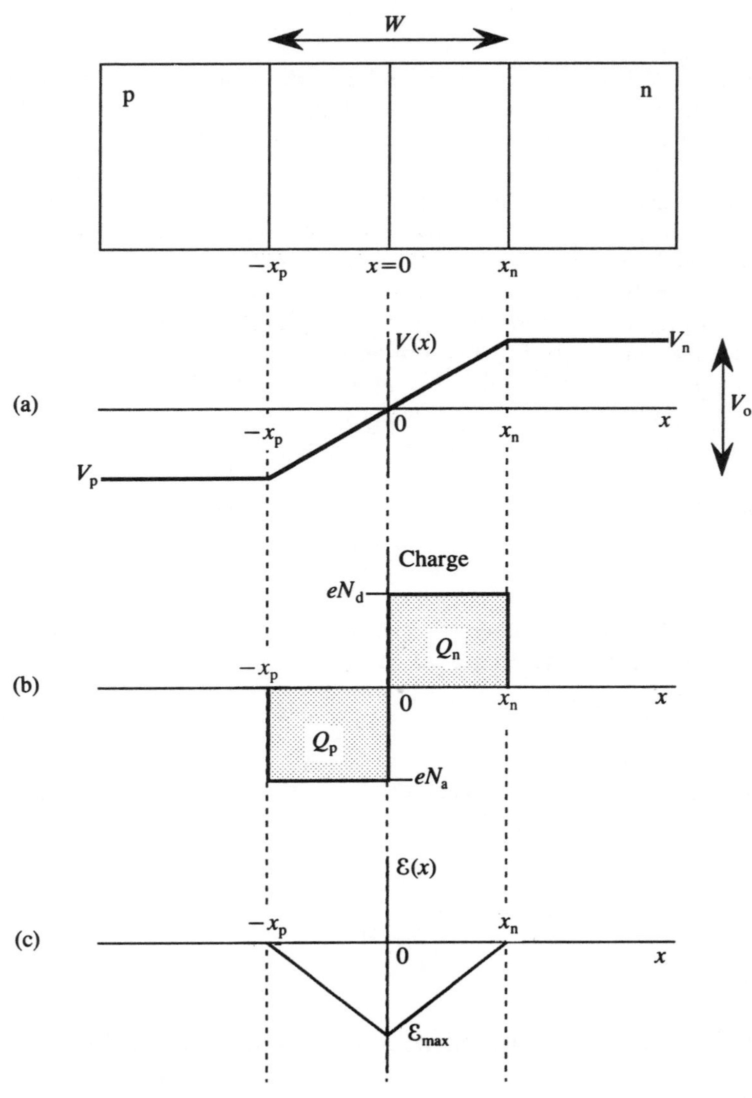

Figure 5.21 *Distributions across a symmetrical abrupt p–n junction: (a) voltage, (b) charge, (c) electric field.*

$$\epsilon_s = \epsilon_o \epsilon_r \tag{5.20}$$

Equations (5.18) and (5.19) indicate that the electric-field distribution on the p-side of the junction has a negative slope and that the distribution on the n-side has a positive slope. These are shown in Figure 5.21(c).

You may prefer to visualize equation (5.17) in its integral form for $\mathcal{E}(x)$:

$$\mathcal{E}(x) = \frac{e}{\epsilon_s} \int (N_d - N_a) \mathrm{d}x \tag{5.21}$$

This means that $\mathcal{E}(x)$ can be found by determining the area under a graph of N_d–N_a versus x. If you inspect Figure 5.21(b) you should be able to see how (c) is derived using this method: on the left-hand side of graph (b), i.e. for the values of x to the left of $-x_p$, the area under the graph is zero. So $\mathcal{E}$ is zero for x to the left of $-x_p$. As we move to the right of $-x_p$ and add up the area under the graph incrementally, we find that the field $\mathcal{E}$ is negative (because Q_p is negative) and that it increases linearly until $x = 0$ is reached. At $x = 0$ the difference $N_d - N_a$ is a maximum, so at $x = 0$ $\mathcal{E}$ must reach a maximum ($\mathcal{E}_{max}$ in the figure). As we continue integrating by adding up the area under the graph to the right of $x = 0$, the difference $N_d - N_a$ decreases linearly until we reach x_n, when the area under graph (b) is zero once again.

$\mathcal{E}_{max}$ can now be determined directly from equation (5.21) by writing

$$|\mathcal{E}_{max}| = \frac{eN_d x_n}{\epsilon_s} = \frac{eN_a x_p}{\epsilon_s} \tag{5.22}$$

We know from equation (5.2) that electric field is the differential of the voltage:

$$\mathcal{E}(x) = \frac{-\mathrm{d}V(x)}{\mathrm{d}x}$$

therefore $-\mathrm{d}V(x)$ must be the integral of $\mathcal{E}(x)$. We also know that $-\mathrm{d}V(x)$ is the contact potential V_o (Figure 5.21(a)). Hence we can write

$$-\mathrm{d}V(x) = -V_o = \int_{-xp}^{x_n} \mathcal{E}(x)\,\mathrm{d}x \tag{5.23}$$

This integral is the area under the $\mathcal{E}(x)$ graph between $-x_p$ and x_n. Hence the contact potential V_o can be determined directly from the graph of $\mathcal{E}$ versus x (Figure 5.22).

$$\begin{aligned} V_o &= \frac{\mathcal{E}_{max}(x_p + x_n)}{2} \\ &= \frac{\mathcal{E}_{max} W}{2} \end{aligned}$$

5.3 Derivation of an equation for depletion width *W*

Now we have an equation for V_o we can go on to find an equation for the depletion width W. From equation (5.24) we can write

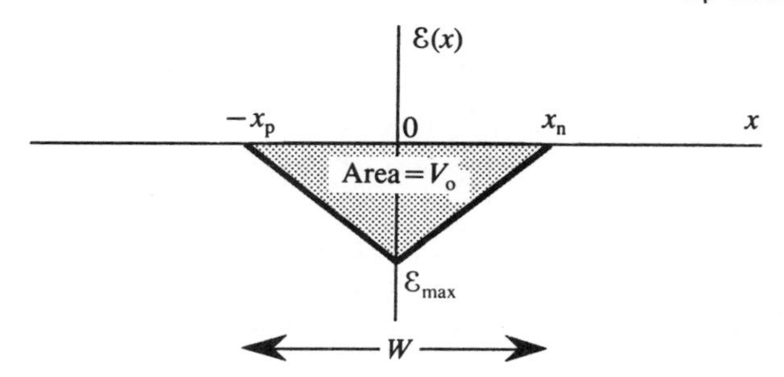

Figure 5.22 *Determination of contact potential V_o from the electric-field distribution across a p–n junction.*

$$V_o = \frac{\mathscr{E}_{max} x_p}{2} + \frac{\mathscr{E}_{max} x_n}{2}$$

Substituting values from equation (5.22) gives

$$V_o = \frac{eN_a x_p x_p}{2\epsilon_s} + \frac{eN_d x_n x_n}{2\epsilon_s}$$

If we tidy this up we get

$$V_o = \frac{e}{2\epsilon_s}(N_a x_p^2 + N_d x_n^2)$$

and replacing $x_p + x_n$ by W leads to

$$V_o = \frac{eW^2}{2\epsilon_s}\left(\frac{N_a N_d}{N_a + N_d}\right)$$

which leads, on rearranging, to:

$$W^2 = \frac{2\epsilon_s V_o}{e}\left(\frac{N_a + N_d}{N_a N_d}\right)$$

$$W = \left[\frac{2\epsilon_s V_o}{e}\left(\frac{N_a + N_d}{N_a N_d}\right)\right]^{1/2} \tag{5.25}$$

This equation for W is important because it's practically useful. N_a and N_d can be measured and ϵ_s is usually known. This gives engineers a means of designing p–n junctions to give a particular value of V_o for a particular value of W.

5.3.1 **Relationship between depletion width and applied bias**

The barrier height in the p–n junction varies with applied bias – it is this phenomenon which produces rectification and is therefore the basis of all p–n diode behaviour. In forward bias the barrier height decreases by the bias voltage V_f and in reverse bias the barrier height increases by V_r (see Section 5.1.3 and Section 5.1.4). This increasing and decreasing barrier height is represented in the equation for depletion width W as follows:

$$W = \left[\frac{2\epsilon_s}{e}\left(\frac{N_a + N_d}{N_a N_d}\right)(V_o - V)\right]^{1/2} \tag{5.26}$$

where V is the applied bias. Note that V must be positive for forward bias so that $(V_o - V)$ is smaller than V_o, and negative for reverse bias so that $(V_o - V)$ is larger than V_o.

5.4 **Asymmetrical abrupt p–n junctions**

In a real p–n junction one side of the junction is usually more heavily doped than the other. This is usually desirable and many devices are designed to exploit the difference in doping, but sophisticated and difficult fabrication processes also mean that a truly symmetrical junction is unlikely. In this section we'll consider an asymmetrical junction which has more dopant atoms on the p-side than on the n-side such that $p > n$.

The depletion region on the p-side of this junction will extend a distance $-x_p$ and the depletion region on the n-side will extend a distance $+x_n$, as previously described (Figure 5.23(a)). For an asymmetrical p–n junction, $|x_p| \neq |x_n|$. (The modulus lines $|x|$ are used here to indicate that we don't need to consider the signs of x_p and x_n.) Recall how the depletion region comes about: diffusing holes recombine with diffusing electrons. If there are many more holes than electrons, as in this case, the electrons diffusing from the n-side into the p-side will be recombined with the holes soon after entering the p-side because there are not many of them, so the depletion region on the p-side will be relatively narrow. On the other hand, there are lots of holes diffusing from the p-side into the n-side, so they will have to penetrate the n-side more deeply to recombine. Hence, when $p > n$, $|x_p| < |x_n|$. The actual metallurgical junction is shown in Figure 5.23 at $x = 0$, as before.

If we know the dopant concentrations either side of the junction we should be able to sketch a graph of the charge-density distribution across the junction. Figure 5.23(b) shows the charge distribution across the asymmetrical p–n junction, where $Q_p = -eN_a A\, x_p$ and $Q_n = eN_d A\, x_n$, where N_a is the acceptor concentration and N_d is the donor concentration. Remember that the depletion width W is given by equation (5.13):

$$W = |x_p| + |x_n|$$

We can also plot the distribution of the built-in electric field $\mathscr{E}$ across W (Figure 5.23(c)).

The one-sided abrupt p–n junction

The extreme example of an asymmetrical p–n junction for which $N_a \gg N_d$ is known as a p^+n junction, the + sign indicating the heavily doped side. Distributions across such a junction are shown in Figure 5.24. Note that if

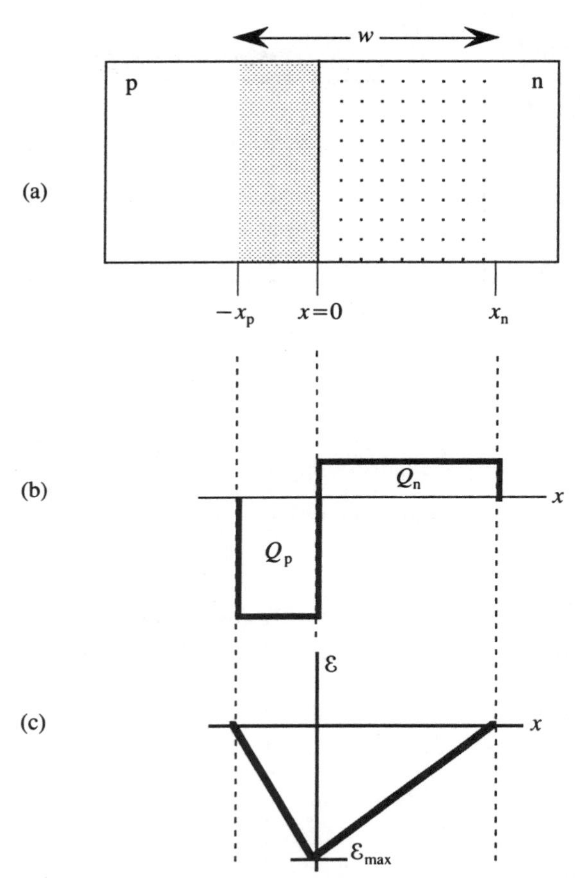

Figure 5.23 *(a) Depletion region in an asymmetrical p–n junction for which $N_a > N_d$, with (b) its charge distribution and (c) its electric-field distribution.*

$$N_a \gg N_d$$

then $|x_p| \ll |x_n|$. Such a junction is called a one-sided abrupt junction.

The penetration of the depletion region is now so small that

$$W \approx x_n$$

in the p^+n junction. The maximum electric field $\mathcal{E}_{max}$ is now given by

$$\mathcal{E}_{max} = \frac{eN_dW}{\epsilon_s} \tag{5.27}$$

SELF-ASSESSMENT QUESTION 5.8

Show that the depletion width in a p^+n junction is given by equation (5.28):

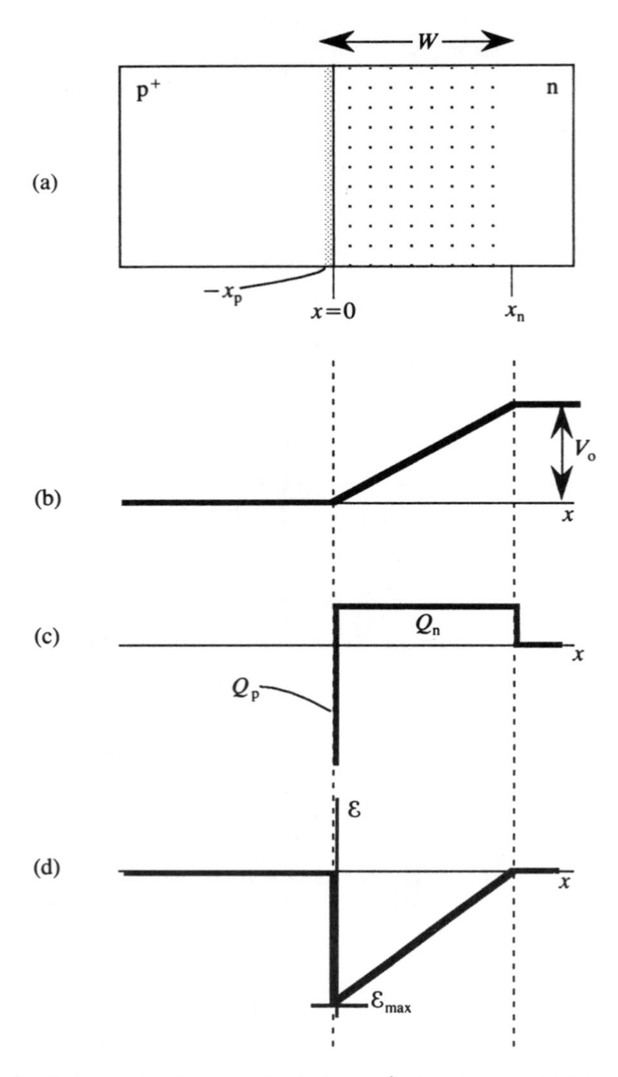

Figure 5.24 *(a) Depletion region in a one-sided abrupt p^+n junction for which $N_a \gg N_d$, with (b) its voltage distribution, (c) its charge distribution and (d) its electric-field distribution.*

$$W \approx x_n = \left[\frac{2\epsilon_s(V_o - V)}{eN_d}\right]^{1/2} \tag{5.28}$$

Similar arguments apply to the pn^+ junction, in which

$$N_a \ll N_d \qquad \text{and} \qquad |x_p| \gg |x_n|$$

SELF-ASSESSMENT QUESTION 5.9

Using your knowledge of p^+n junctions, write an equation for the depletion width W under applied bias V for the pn^+ junction.

5.5 Capacitance of a p–n junction

The depletion layer, by its very nature, will act as a dielectric. Hence there must be a depletion-layer capacitance C_j because the depletion layer of width W acts like a parallel-plate capacitor with a positive charge on one plate and a negative charge on the other (Figure 5.25). Junction capacitance is a useful quantity because its relationship with applied voltage leads to an important experimental technique which enables determination of contact potential V_o and carrier concentration. When designing real semiconductor devices such techniques are invaluable.

5.5.1 Relationship between applied voltage and junction capacitance C_j

The existence of the capacitance provides a very useful way of investigating the depletion width W. They are mathematically related by the following equation:

$$C_j = \epsilon_o \epsilon_r A / W \tag{5.29}$$

where C_j is the junction capacitance in farads, ϵ_o is the permittivity of free space (8.85×10^{-12} F m^{-1}), ϵ_r is the dielectric constant (or relative permittivity) of the semiconductor, A is the cross-sectional area of the junction and W is the depletion width. Note that this equation is often written for unit area, so the A would be left out. Equation (5.29) tells us that a narrow junction (small W) has a larger junction capacitance than a wide one (large W). It follows from this that a forward-biased p–n junction has a larger junction capacitance than one that is reverse-biased.

We can substitute equation (5.26) for W into equation (5.29) to find the following relationship between C_j and applied bias V:

$$C_j = \frac{A}{2}\left[\frac{2e\epsilon_s}{(V_o - V)}\left(\frac{N_a N_d}{N_a + N_d}\right)\right]^{1/2} \tag{5.30}$$

This equation is usually expressed in a slightly different form:

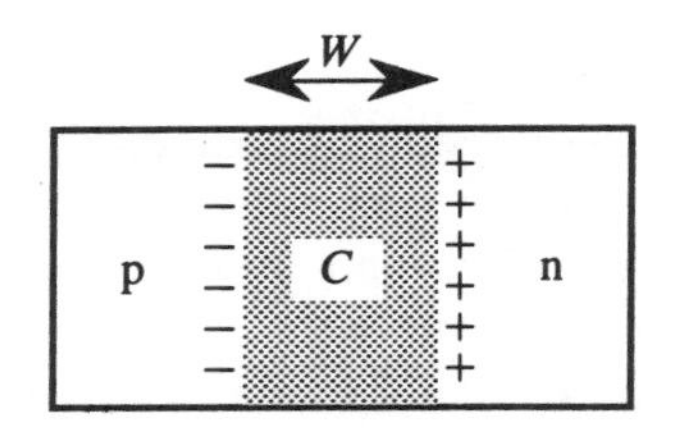

Figure 5.25 *The origin of junction capacitance C_j in a p–n junction.*

$$C_j = A\epsilon_s \left[\frac{e}{2\epsilon_s(V_o - V)} \left(\frac{N_a N_d}{N_a + N_d} \right) \right]^{1/2} \tag{5.31}$$

Junction capacitance therefore varies with the inverse square root of the voltage:

$$C_j \propto 1(V_o - V)^{1/2}$$

5.5.2 The C–V technique for determining barrier height

Equation (5.30) leads to a very useful practical technique for measuring the barrier height V_o (the contact potential). We can rearrange equation (5.30) or (5.31) to give

$$\frac{1}{C_j^2} = \frac{2(V_o - V)}{e\epsilon_o\epsilon_r A^2} \left[\frac{1}{N_a} + \frac{1}{N_d} \right] \tag{5.32}$$

where ϵ_o is the permittivity of free space and ϵ_r is the relative permittivity of the semiconductor. This equation is simplified if we assume that there is a large difference between the sizes of N_a and N_d. For example, if $N_d \gg N_a$, then equation (5.32) becomes

$$\frac{1}{C_j^2} = \frac{2(V_o - V)}{e\epsilon_o\epsilon_r A^2 N_a} \tag{5.33}$$

If a graph is plotted of $1/C_j^2$ versus V a straight line should be obtained (Figure 5.26) if the sample behaviour follows that predicted by theory. Doping concentration N_a (or N_d) and contact potential V_o can be determined from the graph.

SELF-ASSESSMENT QUESTION 5.10

Show how N_a and V_o could both be determined from the sketch graph of Figure 5.26.

5.6 Summary

Electrons and holes move randomly because of their temperature (the warmer they are, the more kinetic energy they'll have, so the faster they move). If an electric field is applied to the sample, the electrons in the sample will drift towards the positive pole. If the sample contains a carrier-concentration gradient, then the carriers in the sample will diffuse to oppose that gradient.

The total current density J in one dimension (the x-direction) is given by the sum of the drift current and the diffusion current, where each type of current consists of two components: hole current and diffusion current. Hence we can write:

$$J_n(x) = e\mu_n n(x)\mathcal{E}(x) + eD_n \frac{dn(x)}{dx}$$

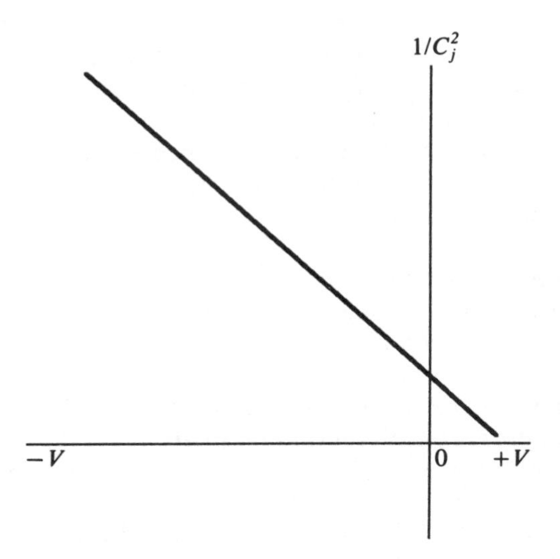

Figure 5.26 *A graph of $1/C_j^2$ versus applied voltage V to determine doping concentration and barrier height V_o.*

where $J_n(x)$ is the electron-current density in the x-direction, and

$$J_p(x) = e\mu_p p(x)\mathcal{E}(x) - eD_p \frac{dp(x)}{dx}$$

where $J_p(x)$ is the hole-current density in the x-direction.

A built-in electric field $\mathcal{E}$ exists across a p–n junction under equilibrium conditions (i.e. zero bias). This process is set up as follows:

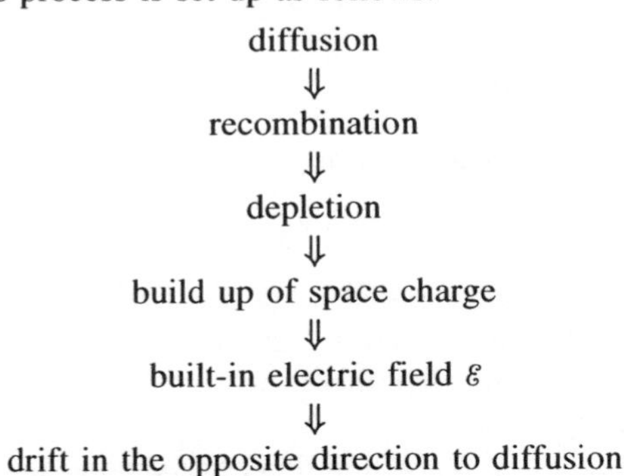

Hence four current components are formed:

- hole diffusion
- hole drift
- electron diffusion
- electron drift

Both the diffusion-current components are in the direction p to n, making up the so-called forward current. Both the drift-current components are in the direction n to p. Under equilibrium conditions the four current components are equal in magnitude, and equations (5.1, 5.4) are obeyed. There is no net current under these conditions.

Under forward bias there are decreases in depletion width W, built-in electric field $\mathcal{E}$, electrostatic-potential barrier height and energy barrier height, and a consequent increase in diffusion current. Drift current remains the same as in the equilibrium case (very small).

Under reverse-bias conditions there are increases in depletion width W, built-in electric field $\mathcal{E}$, electrostatic-potential barrier height and energy barrier height, and a consequent decrease in diffusion current. Drift current remains the same as in the equilibrium and forward-bias case (very small).

The p–n junction exhibits rectification. It conducts a current in forward bias, most of which is due to the diffusion of majority carriers across the depletion width. In reverse bias it conducts a very small current (reverse saturation current I_o) consisting of minority carriers which drift; this is often too small to be measured. At zero bias there is no net current from the device because the diffusion current is equal in magnitude but opposite in direction to the drift current. At zero bias there is a contact potential V_o across the depletion width. The p–n junction is used to make diodes.

The distributions of voltage, charge density and electric field across an abrupt p–n junction have been sketched and equations have been derived relating contact potential, electric field, depletion width, applied bias and doping concentration. Abrupt asymmetrical and one-sided abrupt p–n junctions have been described. Equations have been derived relating junction capacitance to measurable parameters such as applied bias and doping concentration, enabling a practical method for determining barrier height and doping concentration to be described.

5.7 Tutorial questions

5.1 Derive the following equation for yourself.

$$n_n/n_p = \exp(eV_o/kT)$$

5.2 A p–n junction is doped such that it has 5×10^{19} cm^{-3} electrons on the n-side and 2.5×10^{3} cm^{-3} electrons on the p-side. What is the contact potential in this junction? Give the answer in volts *and* electron volts. Assume room temperature.

5.3 An abrupt germanium p–n junction diode has a contact potential of 0.3 V at room temperature. What is the electron concentration on the p-side of the junction? Assume the majority-carrier concentration on the n-side is 2×10^{18} cm^{-3}.

5.4 Draw a fully labelled diagram of each of the following distributions for an asymmetrical abrupt p^{+}n junction:

(i) carrier density (majority and minority)
(ii) electrostatic potential
(iii) charge density
(iv) electric field.

5.5 The relationship between junction capacitance C_j and applied bias V for an abrupt p–n junction is given by the following equation:

$$\frac{1}{C_j^2} = \frac{2(V_o - V)}{e\epsilon_o\epsilon_r A^2}\left[\frac{1}{N_a} + \frac{1}{N_d}\right]$$

where V_o is the contact potential, $\mathcal{E}_r$ is the relative permittivity of the semiconductor, A is the cross-sectional area of the junction, N_a is the acceptor concentration and N_d is the donor concentration. Show how this equation would be modified for a p^+n junction. Use your modified equation to sketch a straight-line graph which would enable doping concentration and contact potential to be determined. Assuming A and $\mathcal{E}_r$ are known, indicate how doping concentration and contact potential would be determined from the graph.

5.6 An abrupt silicon pn^+ junction 10^{-2} cm^2 in area has $N_a = 2 \times 10^{14}$ cm^{-3} doping on the p-side. Calculate the junction capacitance with a reverse bias of 5 V.

5.7 An abrupt silicon p–n junction has $N_a = 5 \times 10^{15}$ cm^{-3} on one side and $N_d = 4 \times 10^{20}$ cm^{-3} on the other. The sample is at room temperature and is kept in the dark. Calculate the Fermi level positions at room temperature in the p and n regions. Draw an equilibrium band diagram for the junction and use the diagram to determine the contact potential V_o. State any assumptions you make.

5.8 The junction described in question (5.7) has a circular cross-section with diameter 100μm. Calculate x_n, x_p, Q_n and $\mathcal{E}_{max}$ for this junction at equilibrium. Sketch the distributions of electric field $\mathcal{E}(x)$ and charge density $Q(x)$ across the junction, marking on the value of the maximum field. Comment on your answers.

5.9 The width of the depletion region of a p–n junction is given by the expression

$$W = \left[\frac{2\epsilon_s}{e}\left(\frac{N_a + N_d}{N_a N_d}\right)V_o\right]^{1/2}$$

How would you modify this expression for depletion width W if you were considering a one-sided abrupt junction of the type pn^+ under zero-bias conditions? State any assumptions you make.

5.10 A silicon pn^+ junction has the electric-field distribution shown in Figure 5.27 under equilibrium conditions.
Calculate the following:

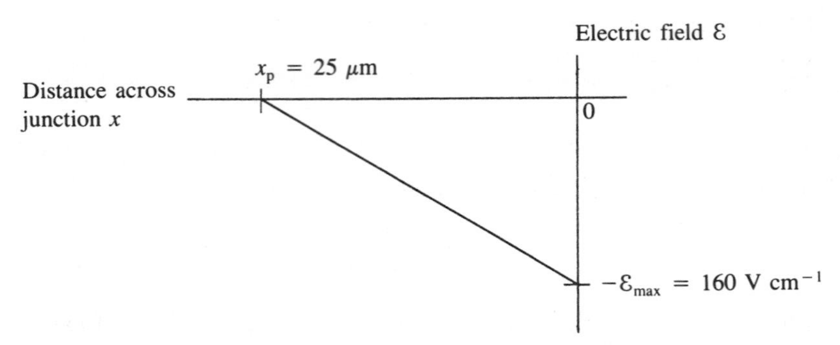

Sketch 5.27 *Sketch graph for tutorial question 5.10.*

(i) contact potential V_o
(ii) acceptor density N_a.
Calculate the junction capacitance for this sample under the following conditions:
(i) forward bias of 0.1 V
(ii) zero bias
(iii) reverse bias of 0.5 V.
Briefly show how you would determine doping density and barrier height using the graph obtained from the so-called *C–V* technique. Roughly sketch a suitable graph using the data obtained to illustrate your answer.

5.11 An abrupt silicon p–n junction diode has a contact potential of 0.7 V at 300 K. What is the hole concentration on the n-side of the junction? Assume the majority-carrier concentration on the p-side is 1×10^{17} cm^{-3}.

5.8 Suggested further reading

Parker, G., The differential form of Gauss's law, in: *Introductory Semiconductor Device Physics*, Prentice Hall, 1994, pp. 76–78.
Streetman, B.G., *Solid State Electronic Devices*, 4th edn, Prentice Hall, 1995, pp. 130–132.

CHAPTER 6

Electrons and semiconductors revisited

Aims and objectives

This chapter develops further the semiconductor models described in Chapters 4 and 5 and introduces some more complex concepts such as wave–particle duality and effective mass. The energy-band diagram is drawn in a new way to show electron momentum, and these diagrams will illustrate that there are two types of bandgap: direct and indirect. A system of statistics called Fermi–Dirac statistics will be introduced to model the behaviour of large numbers of electrons. These statistics will be used to derive equations for electron and hole concentrations in a semiconductor at thermal equilibrium. Intrinsic Fermi level will be defined and the concept of intrinsic carrier concentration will be explained. Equations describing the temperature dependence of carrier concentration will be derived.

6.1 Wave–particle duality

Up until now we've assumed the electron behaves like a particle such as a small sphere. This model has been used because it's convenient — we don't know if electrons are small spheres or not. Sometimes it's more convenient if we model the electron as a wave. There is evidence to support both models, and we should briefly consider this evidence before moving on.

Modern inventions such as the electron microscope provide evidence of the wave nature of electrons. In an electron microscope the source of the 'light' used for 'illumination' of the specimen is an electron gun which fires electrons at the specimen. The electrons in the microscope behave just like light in an ordinary optical microscope — they can be refracted and reflected. In the electron microscope electromagnets are used as lenses and the light 'rays' are streams of electrons. The image produced in the electron microscope could only be produced if the streams of electrons were capable of being refracted and focused, as though they were light waves. So in this example the electron is behaving just like a wave. Another important application of the wave-like properties of the electron is in the electron-diffraction methods used by crystallographers: a stream of electrons will be diffracted by a crystal in the same way that light is diffracted by a crystal. In the semiconductor industry electrons are used instead of ultra-violet light to 'draw' the device patterns on the semiconductor wafers because the short wavelength of the electron enables a higher density of devices to be drawn on a chip. This process is called electron-beam lithography.

Hence we know that electrons may be modelled as waves.

On the other hand, there are some properties of light which can be explained if modelled as particles. Examples include the discrete nature (quantization) of light which is so clearly illustrated by the photoelectric effect. This effect demonstrates that light can be considered as particulate by showing that photons exist.

Hence we can model electrons as particles or as waves.

6.1.1 De Broglie waves and electron momentum

De Broglie was a French physicist (1892–1987) who is acknowledged as the discoverer of the wave nature of particles. He was awarded the Nobel Prize for Physics in 1929. First consider what we mean by momentum p: it's the product of mass m and velocity v.

$$p = mv \tag{6.1}$$

Any moving body has a non-zero momentum. For example, when a person falls down the stairs and breaks an arm, it's the momentum that is responsible for the fracture, not just the velocity or just the mass. A small cat falling down the same stairs at the same speed, for example, will not break a bone because its momentum is small. Likewise another person of the same mass will not break a bone if she or he falls down the stairs much more slowly. Generally, we regard moving objects as having momenta, but what about waves?

Consider Einstein's famous equation

$$E = mc^2 \tag{6.2}$$

where c is the speed of light in vacuo, m is the mass of a body, and E is the energy of that body. This equation was very important because it implied a link between energy E and mass m. On the basis of this equation we can say that as light has associated energy $E = hf$ then that light energy can be associated with a mass. Note, however, that a photon does not have a rest mass. Rearrange equation (6.2) to give:

$$m = \frac{E}{c^2} \tag{6.3}$$

and write

$$E = hf \tag{6.4}$$

for a photon.

Substitute equation (6.4) into equation (6.3) to give

$$m = \frac{hf}{c^2} \tag{6.5}$$

where c is the velocity of the photon (i.e. the velocity of light).

Hence we can write an equation for the momentum of the photon:

$$p = mc$$
$$= \frac{hfc}{c^2} \qquad (6.6)$$

Hence

$$p = \frac{hf}{c} \qquad (6.7)$$

We know that, for a wave, the wavelength λ is given by

$$\lambda = \frac{c}{f}$$

Therefore

$$p = \frac{h}{\lambda} \qquad (6.8)$$

This is the de Broglie equation. It's important because it shows the link between particles (with momentum p) and waves (with wavelength λ). A quantity can now be defined called the de Broglie wavelength λ_{dBr}:

$$\lambda_{dBr} = \frac{h}{p} = \frac{h}{m_r v} \qquad (6.9)$$

where v is now any velocity (it doesn't have to be the velocity of light *in vacuo* c). This equation applies to both particles and waves. It shows that the greater the particle's momentum p, the shorter its de Broglie wavelength λ_{dBr}. Note that, in this equation, the mass m_r is the so-called relativistic mass given by

$$m_r = \frac{m_o}{(1 - v^2/c^2)^{1/2}} \qquad (6.10)$$

which is used when the velocity v of a moving body approaches the speed of light c. Equation (6.10) therefore shows that a body moving at speed c will have an infinite mass m_r. This is one of the strands of Einstein's theory of relativity that illustrates why bodies cannot move at the speed of light — it is assumed that it's impossible for a body to have an infinite mass.

In fact we don't need to consider relativistic speeds because electrons move too slowly in solids for relativistic effects to be taken into account. However, equation (6.8) establishes the link between particles of momentum p and waves of wavelength λ. This implies that electrons may be modelled as waves. Many textbooks include a wave treatment of electrons, and I've included some relevant suggested reading at the end of this chapter. We've also established the principle that the mass of a moving body could be different to its mass when stationary. Our own everyday experience is contrary to this, of course, so this is a very new and startling idea. Einstein's theory of relativity is only relevant because it introduces this idea of changing mass — remember it doesn't apply to electrons in solids, but I hope you can see that Einstein's bold assertion that mass could change does enable the rest of us to consider this in other situations.

6.2 Effective mass

The electrical properties of the lattice should be considered in a discussion of carrier-transport processes. The lattice is important because it has a periodically varying potential caused by the electrical characteristics of the atoms and ions in the lattice. The electrons in the lattice are not completely free to wander around — for instance, they are constrained by having to stay within the confines of the material, unless they become so energetic that they leave the material with a finite kinetic energy and enter the continuum of positive electron energies (as in the photoelectric effect). Also, they cannot be in the same place as other electrons or nuclei at the same time. The electrons react with the periodic potential of the lattice because of their negative charge. The mass of the electron is effectively changed when the electron is moving around the lattice, and its mass will depend on the direction in which it is travelling. Hence a mobile electron has an effective mass m^* which is different to its rest mass m_o. An electron moving at ordinary speeds in a lattice will have an effective mass which depends on how it moves — in which crystallographic direction, for example. The effective mass of an electron is expressed in terms of electron rest mass m_o (Table 6.1). Sometimes different effective masses to those shown are used because those shown in Table 6.1 are *average* values (i.e. averaged over three orthogonal directions x, y, z).

SELF-ASSESSMENT QUESTION 6.1

Using Table 6.1, calculate the effective mass of a hole in gallium arsenide. Give your answer in kilograms to two significant figures.

Table 6.1 *Electron and hole effective masses of some semiconductors at room temperature*

Semiconductor	m_n^*	m_p^*
Germanium Ge	$0.55\ m_o$	$0.37\ m_o$
Silicon Si	$1.1\ m_o$	$0.56\ m_o$
Gallium arsenide GaAs	$0.0067\ m_o$	$0.48\ m_o$
Lead telluride PbTe	$0.17\ m_o$	$0.20\ m_o$
Diamond C	$0.20\ m_o$	$0.25\ m_o$
Indium phosphide InP	$0.077\ m_o$	$0.64\ m_o$

(*Sources*: Streetman, B.G., *Solid State Electronic Devices*, 4th edn, Prentice Hall, 1995; Sze, S.M., *Physics of Semiconductor Devices*, 2nd edn, Wiley, 1981.)

Both carrier types have effective masses, as indicated in Table 6.1, where m_n^* is electron effective mass and m_p^* is hole effective mass.

6.3 *E–k* space and the *E–k* form of the band diagram

$\boldsymbol{k}$ is a quantity called the wave vector. It's associated with electron momentum $\boldsymbol{p}$ and the wave-like behaviour of electrons. Recall the de Broglie wavelength λ_{dBr} (equation (6.9)):

$$\lambda_{\mathrm{dBr}} = \frac{h}{\boldsymbol{p}}$$

where h is Planck's constant.

Now define

$$\lambda_{\mathrm{dBr}} = \frac{2\pi}{\boldsymbol{k}} \tag{6.11}$$

We can now write

$$\frac{h}{\boldsymbol{p}} = \frac{2\pi}{\boldsymbol{k}}$$

therefore

$$\boldsymbol{k} = \frac{2\pi \boldsymbol{p}}{h} \quad \text{or} \quad \boldsymbol{p} = \frac{h}{2\pi}\boldsymbol{k} \tag{6.12}$$

Note that both electron momentum $\boldsymbol{p}$ and the wave vector $\boldsymbol{k}$ are vectors, indicated by the use of bold type. The momentum is therefore directly related to the wave vector by $h/2\pi$.

A commonly used quantity is the so-called 'reduced' form of Planck's constant, given by

$$\hbar = \frac{h}{2\pi} \tag{6.13}$$

therefore

$$\boldsymbol{p} = \hbar \boldsymbol{k} \tag{6.14}$$

Sometimes energy-band diagrams are shown as plots of electron energy E versus $\boldsymbol{k}$ instead of E versus the distance through the crystal. We call this type of space '$\boldsymbol{k}$-space' or 'momentum space'. Note that we plot E against $\boldsymbol{k}$ rather than $\boldsymbol{p}$ just for the sake of convenience. *E–k* diagrams look quite different to the band diagrams you've seen so far. Figure 6.1 shows electron energy E versus $\boldsymbol{k}$ for silicon and for gallium arsenide. Note the positions of the peaks (maxima) and the troughs (minima). Holes in

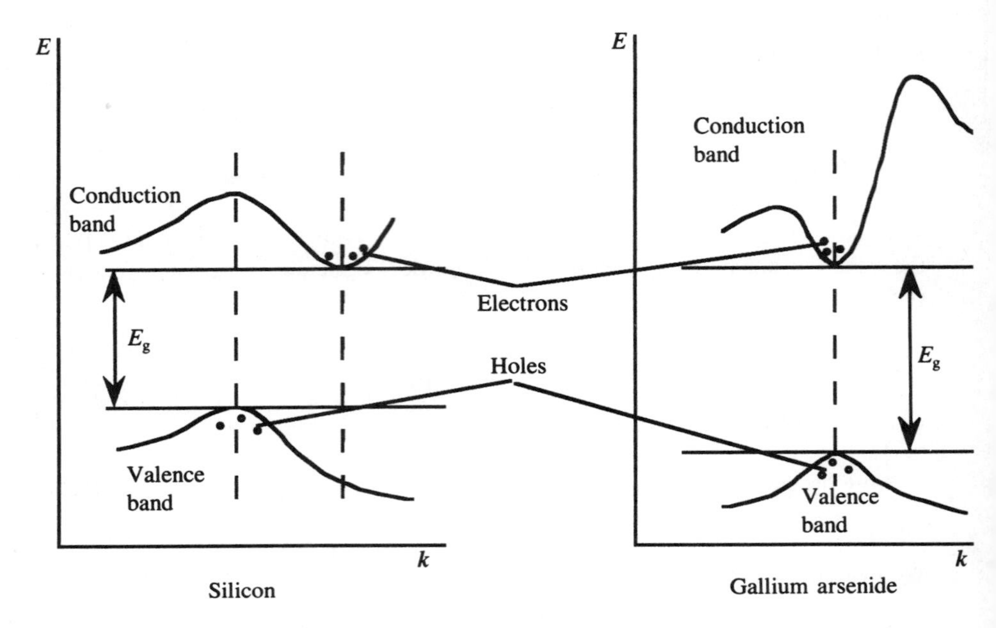

Figure 6.1 ***k*-space band diagrams for silicon and gallium arsenide.**

the valence band lie at the top of the valence band (i.e. in the maximum of the valence band) and electrons in the conduction band lie in the bottom of the conduction band (i.e. in the minimum of the conduction band).

6.3.1 Determination of effective mass from the E–k diagram

The form of the E–$\mathbf{k}$ diagram gives a means of determining effective mass.

Consider Figure 6.2, an E–$\mathbf{k}$ diagram for any semiconductor.

The slope of the graph is $\mathrm{d}E/\mathrm{d}\mathbf{k}$. We know that electron momentum is given by

$$\begin{aligned} \mathbf{p} &= mv \\ &= \hbar\mathbf{k} \end{aligned} \tag{6.14}$$

The kinetic energy of the electron is given by

$$E = \tfrac{1}{2}mv^2 \tag{6.15}$$

where v is the electron's velocity and m its mass.

Hence

$$\begin{aligned} E &= \frac{1}{2}\frac{\mathbf{p}^2}{m} \\ &= \frac{\hbar^2}{2}\frac{\mathbf{k}^2}{m} \end{aligned} \tag{6.16}$$

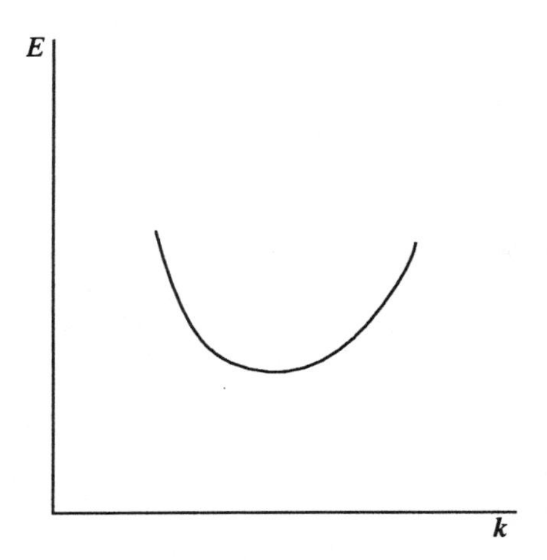

Figure 6.2 *An E–k diagram for a semiconductor.*

Thus electron energy E depends on the square of the wave vector $\boldsymbol{k}$, and the electron mass m is inversely related to the curvature (second derivative[1]) of the E–$\boldsymbol{k}$ relationship. That is, if

$$E = \frac{\hbar}{2}\frac{\boldsymbol{k}^2}{m}$$

then

$$\frac{d^2E}{d\boldsymbol{k}^2} = \frac{d}{d\boldsymbol{k}}\left[\frac{dE}{d\boldsymbol{k}}\right] = \frac{d}{d\boldsymbol{k}}\frac{2\hbar^2\boldsymbol{k}}{2m}$$

$$= \frac{d}{d\boldsymbol{k}}\frac{\hbar^2\boldsymbol{k}}{m}$$

So

$$\frac{d^2E}{d\boldsymbol{k}^2} = \frac{\hbar^2}{m} \qquad (6.17)$$

The slope of the E–$\boldsymbol{k}$ diagram is $dE/d\boldsymbol{k}$, therefore equation (6.17) is the derivative of the slope:

$$\frac{d^2E}{d\boldsymbol{k}^2} = \frac{d\text{ slope}}{d\boldsymbol{k}} \qquad (6.18)$$

1 The second derivative is the derivative of the first derivative, i.e.

$$\frac{d^2y}{dx^2} = \frac{d}{dx}\left(\frac{dy}{dx}\right)$$

The mass of the moving electron in the solid is effective mass m^*, so we can rearrange equation (6.17) to write

$$m^* = \frac{\hbar^2}{d^2E/d\boldsymbol{k}^2} \tag{6.19}$$

Hence it is possible to determine m^* from the slope of the E–$\boldsymbol{k}$ diagram.

From equation (6.19) effective mass may also be expressed in terms of electron momentum $\boldsymbol{p}$:

$$m^* = \frac{d\boldsymbol{p}^2}{d^2E} \tag{6.20}$$

6.4 Direct and indirect semiconductors

In gallium arsenide, the trough of the conduction band lies directly over the peak of the valence band in the E–k diagram (Figure 6.1). A bandgap transition therefore requires no change in k (or momentum) to occur. For example, to excite an electron from the valence band an amount of energy equal to E_g is required, but no momentum is required. If an electron in the conduction band were to recombine with a hole in the valence band, there would be no change in momentum. The situation is different for silicon. For an electron to be excited from the valence band to the conduction band, it not only has to gain energy E_g but also has to gain some momentum, because the trough in the conduction band is not directly over the peak in the valence band on the E–k diagram. When an electron in the conduction band recombines with a hole in the valence band the electron has to lose energy and it has to lose momentum. This momentum is given up as heat to the lattice. Hence gallium arsenide is an example of a direct-bandgap semiconductor and silicon is an example of an indirect-bandgap semiconductor.

Other direct semiconductors include the quaternary compound gallium indium arsenide phosphide GaInAsP, zinc sulphide ZnS and cadmium telluride CdTe. As direct transitions don't require a momentum change to excite electrons into the conduction band or to allow recombination, these transitions are highly efficient. Optoelectronic devices such as light-emitting diodes and laser diodes are made out of direct-gap semiconductors because recombination across a direct bandgap results in the emission of a photon of energy $E_{ph} = E_g$. This happens in exactly the same say that electrons falling from one energy level to a lower one in a Bohr atom emit photons.

Other indirect-gap semiconductors include germanium, Ge, gallium phosphide GaP, and lead telluride, PbTe. Generation and recombination of electron–hole pairs in these materials are inefficient processes because of the heat changes in the lattice and therefore indirect-gap materials are not used to make light-emitting devices.

6.4.1 *Variation of energy-band structure with alloy composition*

A useful property of many semiconductors is their ability to be alloyed together to form ternary and quaternary compound semiconductors. By varying the relative composition of the elements in the alloy it's possible to accurately design the size of the bandgap E_g and sometimes even the nature of the bandgap. Table 6.2 lists some bandgap values for several semiconductors.

From Table 6.2 it can be seen that ternary compounds of GaP and InP will all be of the form $Ga_xIn_{1-x}P$, where x can vary from 0 (i.e. InP) to 1 (i.e. GaP). When$x$ is 0, the bandgap is 1.35 eV, but when x is 1 the bandgap is inceased to 2.26 eV. This means that the size of the bandgap in GaInP can be engineered to be any value between 1.35 eV and 2.26 eV. It can also be seen from Table 6.2 that the quaternary alloy $Ga_xIn_{1-x}As_y P_{1-y}$ can vary in bandgap from 1.35 eV (for $x=0, y=0$) to 1.43 eV (for $x=1, y=1$). This ability to engineer the size of the bandgap is vitally important when designing electronic devices.

Table 6.2 also shows that the bandgap of some alloys may be direct or indirect, depending on alloy composition. $Al_{0.3}Ga_{0.7}As$ has a direct bandgap of 1.87 eV but if the aluminium:gallium ratio is increased to 0.4:0.6 the bandgap becomes indirect and 1.96 eV.

SELF-ASSESSMENT QUESTION 6.2

Use the information in Table 6.2 to determine the wavelengths of the light emitted in $Al_{0.3}Ga_{0.7}As$ and InP when electrons recombine across the bandgap. Give your answers in μm. Which of these emissions would be visible to the human eye, assuming sufficient emission intensity?

Table 6.2 *The nature and size of some semiconductor bandgaps at room temperature*

Semiconductor		*Bandgap type*	*Bandgap(eV)*
Indium arsenide	InAs	Direct	0.36
Germanium	Ge	Indirect	0.67
Silicon	Si	Indirect	1.11
Indium phosphide	InP	Direct	1.35
Gallium arsenide	GaAs	Direct	1.43
Aluminium gallium arsenide	$Al_{0.3}Ga_{0.7}As$	Direct	1.87
Gallium arsenide phosphide	$GaAs_{0.6}P_{0.4}$	Direct	1.94
Aluminium gallium arsenide	$Al_{0.4}Ga_{0.6}As$	Indirect	1.96
Gallium arsenide phosphide	$GaAs_{0.5}P_{0.5}$	Indirect	2.03
Aluminium arsenide	AlAs	Indirect	2.16
Gallium phosphide	GaP	Indirect	2.26
Gallium indium arsenide phosphide	GaInAsP	Direct–Indirect	0.36–2.26
Zinc sulphide	ZnS	Direct	3.68
Diamond	C	Indirect	5.5
Diamond	C	Direct	7.3

6.5 Fermi–Dirac statistics

6.5.1 *Systems of statistics*

Engineers use models to predict the behaviour of real systems. When the system is a large population of items, it may be necessary to treat the population statistically in order to predict its behaviour. There are some important physical systems which are modelled using statistics, such as populations of alpha particles or electrons. For example, in a $1\,\mathrm{cm}^3$ sample of semicondutor there could be more than 1×10^{15} electrons in the conduction band. That's a very large number to deal with, and modelling the behaviour of every individual electron in the population would be extremely difficult. Using statistics is a viable alternative.

There are three well-known systems of statistics used by scientists and engineers: Fermi–Dirac statistics, Bose–Einstein statistics, and Maxwell–Boltzmann statistics. Each of these systems of statistics is used to model a particular type of population. In the case of Fermi–Dirac statistics, the population must be made up of indistinguishable items called fermions. The most famous fermion is the electron. A description of all these systems of statistics is beyond the scope of this book, but there is some suggested reading on the subject at the end of the chapter.

6.5.2 *The Fermi–Dirac distribution function*

The application of Fermi–Dirac statistics to the case of electrons in solids leads to a useful equation which gives the probability of an energy level E being occupied by an electron. This probability is called the Fermi–Dirac distribution function $f(E)$. It's a distribution because plotting $f(E)$ versus E gives a picture of how electrons are distributed as a function of electron energy E. The Fermi–Dirac distribution function $f(E)$ is usually expressed in the form of an equation:

$$f(E) = \frac{1}{1+\exp[(E-E_\mathrm{f})/kT]} \tag{6.21}$$

where $f(E)$ is the probability of finding an electron in an energy level E at temperature T. The energy level E_f is the Fermi level of the material containing the energy level E. The distribution function $f(E)$ is plotted against E in Figure 6.3, for three different temperatures: 0 K, 300 K and 1500 K. The shapes of the resulting curves are interesting so we'll consider them further.

Temperature dependence

Let's examine what happens in a semiconductor at 0 K. At this temperature the valence band is full of electrons and the conduction band is empty (Figure 4.2). Therefore we might expect that the Fermi–Dirac distribution function would tell us that, at 0 K, there is zero probability of finding an electron at an energy level higher than E_f and 100% probability of finding an electron at an energy level lower than E_f.

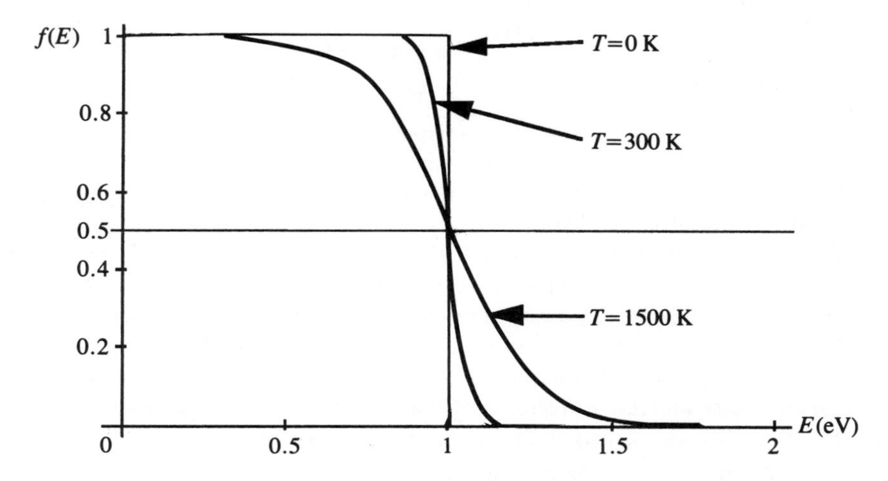

Figure 6.3 *Fermi–Dirac distribution function for T = 0 K, T = 300 K and T = 1500 K (for E_f = 1.0 eV).*

Inspection of Figure 6.3 shows that this is the case, and at 0 K $f(E)$ is square. This can easily be shown mathematically, as follows.

At T = 0 K,

$$f(E) = \frac{1}{1 + \exp[(E - E_f)/0]}$$

We can consider three cases at T = 0 K: $E < E_f$, $E = E_f$, $E > E_f$.

1. $E < E_f$

$$f(E) = \frac{1}{1 + \exp[-\infty]}$$

$$= \frac{1}{1 + 0} = 1$$

Hence there is 100% probability of finding an electron at an energy level E which is lower than E_f at absolute zero.

2. $E = E_f$

$$f(E) = \frac{1}{1 + \exp[0]}$$

$$= \frac{1}{1 + 1} = 0.5$$

Hence the probability of finding an electron at an energy level E which is at E_f at absolute zero is a half.

3. $E > E_f$

$$f(E) = \frac{1}{1+\exp[+\infty]} = \frac{1}{1+\infty} = 0$$

Hence there is zero probability of finding an electron at an energy level E which is higher than E_f at absolute zero.

If the temperature of the material is then raised, say to 300 K, some electrons in the valence band will gain enough thermal energy to be excited across the bandgap, thus generating an electron–hole pair (Figure 4.4). Inspection of Figure 6.3 shows that the distribution function at 300 K is no longer square. At even higher temperatures the square shape of the distribution function is lost as many more electron–hole pairs are generated. There are certain features of Figure 6.3 which should be noted:

- all curves pass through the same point $f(E) = 0.5$ at $E = E_f$;
- the shape of each curve is the same either side of $E = E_f$;
- $f(E)$ lies between 0 and 1;
- the area under each curve is the same.

6.6 Determination of carrier concentration

6.6.1 *Density of states and carrier concentration at equilibrium*

$f(E)$ tells us the probability of finding an electron in an energy state E but it doesn't tell us if the energy state at E exists or not. So $f(E)$ on its own isn't very much use to us. However, if we multiply this probability $f(E)$ by the number (or density or concentration) of energy states at energy E, which we'll call $N(E)$, then we get the number of states at E which are likely to be occupied by an electron. In other words, we'll get the density of electrons occupying energy states of energy E. That is,

number of electrons at energy E = probability of finding an electron at E × number of available states at E

= $f(E)\,N(E)$

We can make this more useful by writing an equation for the concentration of electrons in the conduction band at thermal equilibrium:

$$n_o = \int_{E_c}^{\infty} f(E)\,N(E)\,\mathrm{d}E \tag{6.22}$$

where n_o = electron concentration in the conduction band at thermal equilibrium; $f(E)$ = probability of an energy state E being occupied by an electron; $N(E)\mathrm{d}(E)$ = density of available states in the energy range $\mathrm{d}E$. This expression tells us that we can find the

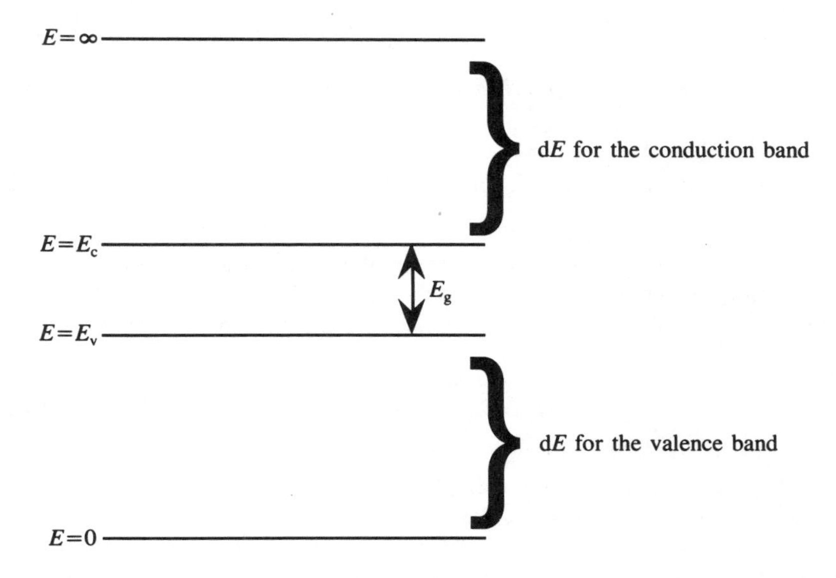

Figure 6.4 *Definition of the energy ranges dE for the conduction and valence bands.*

number of electrons per cubic metre (or per cubic centimetre) in the conduction band by integrating over all the energy states between $E = E_c$ and $E = \infty$, where we define E_c and $E = \infty$ as the boundaries of the conduction band (Figure 6.4).

We can write a similar expression for the number of holes in the valence band at thermal equilibrium. Remember that the probability of finding a hole is equal to one minus the probability of finding an electron. Therefore,

$$p_o = \int_0^{E_v} [1 - f(E)]\, N(E)\, dE \tag{6.23}$$

where p_o = hole concentration in the valence band at thermal equilibrium; $1 - f(E)$ = probability of an energy state E being occupied by a hole; $N(E)\mathrm{d}(E)$ = density of available states in the energy range dE from $E = 0$ to $E = E_v$.

Effective density of states

The function $N(E)$ can be calculated using quantum mechanics and the Pauli exclusion principle. I'm just going to state the result of this. It turns out that the density of states $N(E)$ in the conduction band is equal to the density of states at the conduction-band edge E_c. Similarly, the density of states $N(E)$ in the valence band is equal to the density of states at the valence-band edge E_v. These are called the effective density of states in the conduction band, N_c, and the effective density of states in the valence band, N_v.

Therefore we can write

$$n_o = \int_{E_c}^{\infty} f(E)\,N(E)\,\mathrm{d}E = f(E_c)N_c\ \mathrm{m}^{-3} \tag{6.24}$$

$$p_o = \int_{0}^{E_v} [1 - f(E)]\,N(E)\,\mathrm{d}E = [1 - f(E_v)]N_v\ \mathrm{m}^{-3} \tag{6.25}$$

$f(E_c)$ is the probability of finding an electron at the conduction-band edge E_c; $f(E_v)$ is the probability of finding an electron at the valence band edge E_v.

The effective density of states N_c in the conduction band is given by

$$N_c = 2\left(\frac{2\pi m_n^* kT}{h^2}\right)^{3/2}\ \mathrm{m}^{-3} \tag{6.26}$$

where m_n^* is the electron effective mass, k is Boltzmann's constant, T is the absolute temperature and h is Planck's constant.

Similarly the effective density of states N_v in the valence band is given by

$$N_v = 2\left(\frac{2\pi m_p^* kT}{h^2}\right)^{3/2}\ \mathrm{m}^{-3} \tag{6.27}$$

where m_p^* is the hole effective mass.

We now have the means of calculating the electron concentration and the hole concentration at thermal equilibrium:

$$n_o = f(E_c)2\left(\frac{2\pi m_n^* kT}{h^2}\right)^{3/2}\ \mathrm{m}^{-3} \tag{6.28}$$

$$p_o = [1 - f(E_v)]2\left(\frac{2\pi m_p^* kT}{h^2}\right)^{3/2}\ \mathrm{m}^{-3} \tag{6.29}$$

SELF-ASSESSMENT QUESTION 6.3

What is the probability of finding an electron at the conduction-band edge in a sample of silicon at room temperature? Assume the conduction-band edge is 0.01 eV above the Fermi level. Give your answer to two significant figures.

SELF-ASSESSMENT QUESTION 6.4

What is the effective density of states at the conduction-band edge in this silicon sample? Assume the effective mass of electrons in silicon is 1.1 m_o.

Give your answer in m^{-3} and cm^{-3}. Give your answer to two significant figures.

SELF-ASSESSMENT QUESTION 6.5

How many electrons are there in the conduction band of this silicon sample at room temperature? Give your answer to two significant figures.

SELF-ASSESSMENT QUESTION 6.6

Would you say this silicon sample was intrinsic or extrinsic? If the latter, would you say it was n-type or p-type? Explain your answer.

6.6.2 Intrinsic Fermi level E_i

The electron and hole concentrations n_o and p_o obtained from our equations are valid whether the material is intrinsic or extrinsic, provided thermal equilibrium is maintained. We know, however, that for intrinsic material E_f lies close to the middle of the bandgap. This intrinsic Fermi level is useful, and is given its own symbol E_i. The number of electrons in an intrinsic material can therefore be given by:

$$
\begin{aligned}
n_i &= N_c f(E_c) \\
&= N_c \frac{1}{1 + e^{(E_c - E_i)/kT}} \qquad (6.30)
\end{aligned}
$$

Likewise, the number of holes in an intrinsic material is given by

$$
\begin{aligned}
p_i &= N_v[1 - f(E_v)] \\
&= N_v \left[1 - \frac{1}{1 + e^{(E_v - E_i)/kT}}\right] \qquad (6.31)
\end{aligned}
$$

It would be useful to tidy up these expressions to make them easier to use. For instance, consider the 1 in the denominator of the Fermi–Dirac distribution function:

$$
f(E) = \frac{1}{1 + e^{(E - E_f)/kT}} \qquad (6.21)
$$

Do we need to keep this 1 in the denominator? If we showed that $e^{(E_c - E_f)/kT} \gg 1$ at the temperatures we're interested in, then the 1 in the denominator can be neglected. We can assume that, for an intrinsic semiconductor, the Fermi level lies at least several kT below the conduction band. This makes the exponential term large compared to 1. For example, consider intrinsic silicon at room temperature.

Sketch a band diagram to represent silicon at room temperature (Figure 6.5). The bandgap is 1.1 eV. Assume the intrinsic fermi level E_i lies halfway in the bandgap. Therefore,

$$E_c - E_i = 0.55 \text{ eV}$$

hence

$$(E_c - E_i)/kT = 0.55/0.025 = 22$$

$$e^{22} = 3.6 \times 10^9$$

Hence $e^{(E_c-E_i)/kT} \gg 1$ at room temperature, showing that we can confidently neglect the 1 in the denominator of our expression for n_i. So let's rewrite our expression for n_i:

$$n_i = N_c \frac{1}{e^{(E_c-E_i)/kT}} = N_c[e^{(E_c-E_i)/kT}]^{-1}$$

$$= N_c e^{-(E_c-E_i)/kT} = N_c e^{(E_i-E_c)/kT} \tag{6.32}$$

We can assume that the Fermi level lies at least several kT above the valence band. Consider intrinsic silicon at room temperature again.

$$E_v - E_i = -0.55 \text{ eV}$$

hence

$$(E_v - E_i)/kT = -0.55/0.025 = -22$$

$$e^{22} = 2.8 \times 10^{-10}$$

This time the exponential term is extremely small, such that $1 \gg e^{(E_v-E_i)/kT}$.

Consider the probability of finding a hole at E_v:

$$1 - f(E_v) = 1 - \frac{1}{1 + e^{(E_v-E_i)/kT}} \tag{6.33}$$

Find a common denominator:

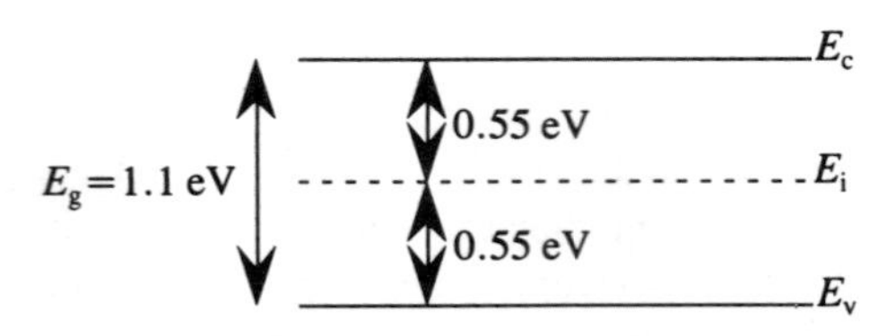

Figure 6.5 *Simple band diagram representing intrinsic silicon at room temperature.*

$$1 - f(E_v) = \frac{1 + e^{(E_v - E_i)/kT} - 1}{1 + e^{(E_v - E_i)/kT}}$$

$$= \frac{e^{(E_v - E_i)/kT}}{1 + e^{(E_v - E_i)/kT}}$$

but we know that $e^{(E_v - E_i)/kT} \ll 1$, therefore the exponential term in the denominator can be neglected. This gives

$$1 - f(E_v) = \frac{e^{(E_v - E_i)/kT}}{1}$$

$$= e^{(E_v - E_i)/kT} \tag{6.34}$$

Therefore we can find the density p_i of holes in an intrinsic material:

$$p_i = N_v e^{(E_v - E_i)/kT} \qquad \text{or} \qquad p_i = N_v e^{-(E_i - E_v)/kT} \tag{6.35}$$

6.6.3 Intrinsic carrier concentration n_i

In an intrinsic material the concentration of electrons in the conduction band is equal to the concentration of holes in the valence band.

$$n_i = p_i \tag{6.36}$$

Now let's find an expression for $n_o p_o$, assuming that the conduction-band edge is several kT higher than the Fermi level E_f and assuming that the valence-band edge is several kT below the Fermi level E_f. Therefore,

$$n_o = N_c e^{-(E_c - E_f)/kT}$$

$$p_o = N_v e^{-(E_f - E_v)/kT}$$

Therefore

$$n_o p_o = N_c e^{-(E_c - E_f)/kT} N_v e^{-(E_f - E_v)/kT}$$

$$= N_c N_v e^{[-(E_c - E_f)/kT] + [-(E_f - E_v)/kT]}$$

$$= N_c N_v e^{[-E_c + E_f - E_f + E_v]/kT}$$

$$= N_c N_v e^{-E_g/kT} \tag{6.37}$$

where E_g is the material's bandgap.

Now consider the product $n_i p_i$.

$$n_{\mathrm{i}} = N_{\mathrm{c}}\mathrm{e}^{-(E_{\mathrm{c}}-E_{\mathrm{i}})/kT}$$

$$p_{\mathrm{i}} = N_{\mathrm{v}}\mathrm{e}^{-(E_{\mathrm{i}}-E_{\mathrm{v}})/kT}$$

Therefore

$$n_{\mathrm{i}}p_{i} = N_{\mathrm{c}}N_{\mathrm{v}}\mathrm{e}^{-E_{\mathrm{g}}/kT} \tag{6.38}$$

However,

$$n_{\mathrm{i}} = p_{\mathrm{i}}$$

Therefore

$$n_{\mathrm{i}}p_{\mathrm{i}} = n_{\mathrm{i}}^2.$$

Hence we can write an expression for n_{i} by writing the square root of equation (6.38):

$$n_{\mathrm{i}} = (N_{\mathrm{c}}N_{\mathrm{v}})^{1/2}\mathrm{e}^{-E_{\mathrm{g}}/2kT} \tag{6.39}$$

Therefore

$$n_{\mathrm{o}}p_{\mathrm{o}} = n_{\mathrm{i}}^2 \tag{6.40}$$

This is an important expression. It shows that the product $n_{\mathrm{o}}p_{\mathrm{o}}$ at thermal equilibrium is a constant for a particular material and temperature, even if the doping is varied. n_{i} is called the intrinsic carrier concentration. For silicon at room temperature n_{i} is approximately $1.5\times10^{10}\,\mathrm{cm}^{-3}$.

SELF-ASSESSMENT QUESTION 6.7
What is $1.5 \times 10^{10}\,\mathrm{cm}^{-3}$ expressed in m^{-3}?

n_{o} and p_{o} can also be written as

$$n_{\mathrm{o}} = n_{\mathrm{i}}\mathrm{e}^{(E_{\mathrm{f}}-E_{\mathrm{i}})/kT} \tag{6.41}$$

$$p_{\mathrm{o}} = n_{\mathrm{i}}\mathrm{e}^{(E_{\mathrm{i}}-E_{\mathrm{f}})/kT} \tag{6.42}$$

These last two equations for n_{o} andp_{o} clearly show that the electron concentration is n_{i} when E_{f} is located at the intrinsic level E_{i}. They also show that n_{o} increases exponentially as the Fermi level moves from E_{i} towards E_{c}, the conduction-band edge (i.e. as $(E_{\mathrm{f}}-E_{\mathrm{i}})$ increases). This is further evidence for the high position of the Fermi level in the bandgap of an n-type material. Similar arguments apply to p-type material.

6.6.4 *Temperature dependence of carrier concentration*

The previous equations show that n_{o}, p_{o} and n_{i} are all temperature dependent. By now you should know that carrier concentration will be affected by temperature because the number of electron–hole pairs generated depends on the temperature of the material. The higher the temperature, the more EHPs are generated. Let's examine this temperature dependence in more detail by considering the intrinsic carrier concentration n_i. From previous work, we know

$$n_i = (N_c N_v)^{1/2} e^{-E_g/2kT} \tag{6.39}$$

Substituting in for N_c and N_v produces

$$n_i = \left[2\left(\frac{2\pi m_n^* kT}{h^2}\right)^{3/2}\right]^{1/2}\left[2\left(\frac{2\pi m_p^* kT}{h^2}\right)^{3/2}\right]^{1/2} e^{-E_g/2kt}$$

$$= \left[2\left(\frac{2\pi kT}{h^2}\right)^{3/2} m_n^{*3/2}\right]^{1/2}\left[2\left(\frac{2\pi kT}{h^2}\right)^{3/2} m_p^{*3/2}\right]^{1/2} e^{-E_g/2kT}$$

$$= 2\left(\frac{2\pi kT}{h^2}\right)^{3/2} (m_n^* m_p^*)^{3/4} e^{-E_g/2kT} \tag{6.43}$$

Note that all the parameters except T are constant for a given material. The temperature T appears twice in this equation: $T^{3/2}$ and $e^{-1/T}$. The most dominant of these is the exponent, i.e. the exponential term in T has the strongest influence on the value of n_i. If we were to assume that the $T^{3/2}$ term is virtually constant compared to the fast-changing $e^{-1/T}$ term, we could write

$$n_i \approx \text{constant} \times e^{-E_g/2kT} \tag{6.44}$$

Taking natural logarithms of both sides leads to

$$\ln n_i = -E_g/2kT \tag{6.45}$$

This is now in the form of a straight-line equation so if we were to plot $\ln n_i$ versus $1/T$ we should get a straight line of slope $-E_g/2k$ (assuming that the exponential term in T is much more dominant than $T^{3/2}$). In fact this plot is almost linear, proving that the change in $e^{-1/T}$ is much more dominant than the change in $T^{3/2}$. Hence it's easy to find n_i for any material at any temperature, just by calculating $\ln n_i$ for different values of T and plotting against $1/T$.

So if we know the temperature of a sample, we can read off a value for n_i. To determine equilibrium electron concentration we then need to know $E_f - E_i$:

$$n_o = n_i e^{(E_f - E_i)/kT}$$

Alternatively, if we want to find the position of the Fermi level E_f relative to E_i we must know n_o.

6.7 **Summary**

The wave nature of electrons has been introduced and the concept of electron momentum has been explained. The *E–k* diagram has been briefly explained and the concept of effective mass introduced. Direct- and indirect-bandgap semiconductors have been described. The importance of the alloy composition of semiconductor compounds has been described, and examples have been given of how alloy composition varies bandgap. Fermi–Dirac statistics have been summarized and used to determine equations for the carrier concentrations in a semiconductor. The effective density of states equations have been stated for the conduction band and for the valence band. Intrinsic carrier concentration n_i has been introduced and shown to be a useful quantity. Equations describing the temperature dependence of carrier concentration have been derived.

6.8 Tutorial questions

6.1 Inspect the equations that have been derived for n_i and p_i and satisfy yourself that the intrinsic Fermi level E_i must lie in the middle of the bandgap if $N_c = N_v$. Note that N_c is unlikely to equal N_v because hole effective mass is unlikely to equal electron effective mass. This means that the intrinsic Fermi level E_i won't lie *exactly* in the middle of the bandgap.

6.2 Consider a sample of gallium arsenide GaAs at room temperature which is doped with 10^{18} atoms of phosphorus P per cubic centimetre. What is the electron concentration n_o at room temperature? Draw the equilibrium band diagram for this sample, showing E_f and E_i on your sketch. Assume the bandgap of gallium arsenide at room temperature is 1.4 eV.

6.3 Plot a graph to show how the Fermi–Dirac distribution function $f(E)$ varies with temperature. Use the temperatures 0 K, 300 K and 1500 K, and plot $f(E)$ versus $(E–E_f)$. Indicate the Fermi level E_f on your graph. Explain the graph's physical significance.

6.4 What is the probability of occupancy by an electron at room temperature in an energy state E 0.05 eV below the Fermi level? Assuming the energy state E is at the top of the valence band, would you say the sample is n-type or p-type, and why? What would the majority-carrier concentration be for this sample? Assume the sample is gallium arsenide GaAs with a hole effective mass of $0.48m_o$. Give your answer in m^{-3} and cm^{-3}.

6.5 The Fermi level in a sample of gallium arsenide is 0.25 eV below the conduction band. What is the probability that energy levels at the bottom of the conduction band will be occupied by an electron when the temperature is 300 K? If the bandgap is 1.42 eV, what is the probability that a level at the top of the valence band will contain a hole? What would the electron concentration be in the conduction band of this sample? Assume the sample has an electron effective mass of $0.067m_o$. Give your answer in m^{-3} and cm^{-3}.

6.6 Explain why gallium arsenide is a more suitable material than silicon for the fabrication of optoelectronic devices such as light-emitting diodes and laser diodes.

6.7 The Fermi level in a sample of silicon is 0.6 eV below the conduction band. What is the probability that energy levels at the top of the valence band will be occupied by an electron at 600 K? Assume the bandgap is 1.1 eV. What is the probability that a level at the bottom of the conduction band will contain a hole at this temperature? Give your answers to seven decimal places and comment on their values. What would the electron carrier concentration be in the conduction band of this sample at 600 K? Assume an electron effective mass in silicon of $1.1m_o$.

6.8 What is the probability of occupancy by an electron at 300 K in an energy state E which is 0.25 eV above the Fermi level? If the energy state E of this sample is at the bottom of the conduction band, what would the electron carrier concentration be? Assume the sample is n-type gallium arsenide with an electron effective mass of $0.067m_0$. Give your answer in m^{-3} *and* cm^{-3}.

6.9 Calculate the following:

- relativistic mass m_r of a body of rest mass m_o which is moving at the velocity of light c;
- relativistic mass m_r of an electron travelling at a velocity of 1×10^5 m s^{-1};
- thermal velocity of an electron at room temperature using equation (4.1).

Comment on your answers.

6.10 Consider the case of the person falling down the stairs. Assume the person's rest mass m_o is 50 kg, and assume that the velocity at which the person falls is 3 m s^{-1}. What is the relativistic mass m_r?

6.11 Show that we do not need to take relativistic speeds into account when considering electrons moving in a crystal lattice.

6.12 Why do you suppose it is reasonable to assume that the effective density of states in the conduction band N_c is equal to the density of states at E_c?

6.9 Suggested further reading

Blakemore, J. S., Appendix A, The Fermi–Dirac distribution law, in: *Semiconductor Statistics*, Dover, 1987, pp. 343–345.

Parker, G., *Introductory Semiconductor Device Physics*, Prentice Hall, 1994, pp. 248–266.

Sears, F. W. and Salinger, G. L., *Thermodynamics, Kinetic Theory and Statistical Thermodynamics*, 3rd edn, Addison-Wesley, 1975, pp. 302–322.

Streetman, B.G., *Solid State Electronic Devices*, 4th edn, Prentice Hall, 1995, pp. 71–80.

CHAPTER 7

Diodes and their operation

Aims and objectives

Earlier chapters described the operation of a p–n junction according to a simple model. This chapter extends this model to diodes, and begins by considering the I–V characteristic of the p–n junction diode. Simple circuits are used to illustrate rectification using a p–n diode and voltage regulation using a zener diode. By the end of this chapter readers should be able to understand the operation of various diode types in terms of the p–n junction model developed earlier.

7.1 The p–n junction diode

Chapter 5 described a simple model of what happens across the depletion region in a p–n junction. We saw how an energy barrier is built into the p–n junction by the transport of carriers across the junction. The barrier is responsible for the ease with which a current crosses the depletion region from the p-side to the n-side when the p-side of the junction is made positive relative to the n-side. The same barrier is also responsible for the difficulty in passing a current when the p-side is made negative relative to the n-side. This ability to pass a current when biased one way and not the other is called rectification.

Figure 7.1 illustrates rectification in a p–n junction. The large current passing from p to n is called the forward current. The way the voltage is applied to the diode to cause the forward current to flow is called forward bias (Figure 7.1(a)). The large current which flows under forward bias is made up of majority carriers diffusing from the p-side to the n-side. When the applied voltage is reversed such that the p-side is negative relative to the n-side (Figure 7.1(b)) a very tiny current flows from the n-side to the p-side. This is called the reverse current and the bias at this time is called the reverse bias. The reverse current is made up of minority carriers drifting from the n-side to the p-side. When reverse-biased the current passing through the diode is very small, typically about 1 nA. The size of the forward current is typically tens of milliamperes or greater.

Diodes are electronic devices which rectify. There are several types of diode, but the most common is the p–n junction diode. All diodes contain an energy barrier at a junction which causes the rectification.

The circuit symbol for all diodes represents the phenomenon of rectification by an arrowhead — a large current is passed in the direction of the arrowhead when the diode

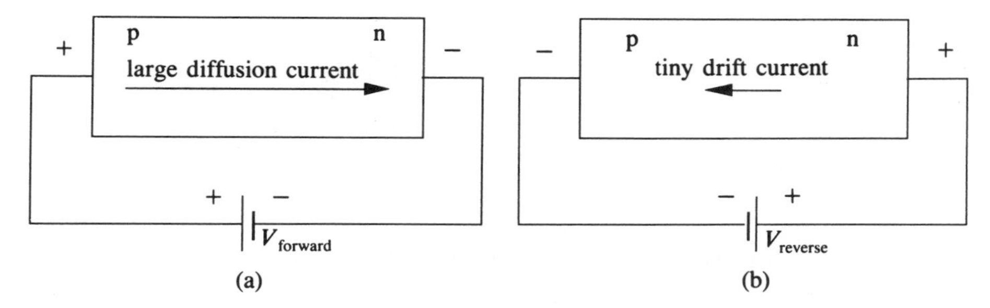

Figure 7.1 *The dependence of current flow in a p–n junction on applied bias: (a) forward bias; (b) reverse bias.*

is forward-biased. Figure 7.2 shows the circuit symbol for an ordinary p–n diode, and Figure 7.3 shows some of the labels used to identify the two diode electrodes. You should always consult the manufacturer's information to identify which electrode is the anode and which is the cathode — often the cathode is identified by a black band at the cathode end of the diode (Figure 7.4).

It's essential that you connect the diode the right way round when building electronic circuits — some diodes are designed to operate in forward bias (e.g. the light-emitting diode) and some are designed to operate in reverse bias (e.g. the zener diode). These different types of diode will be examined later in this chapter.

Figure 7.2 *Circuit symbol for a p–n junction diode.*

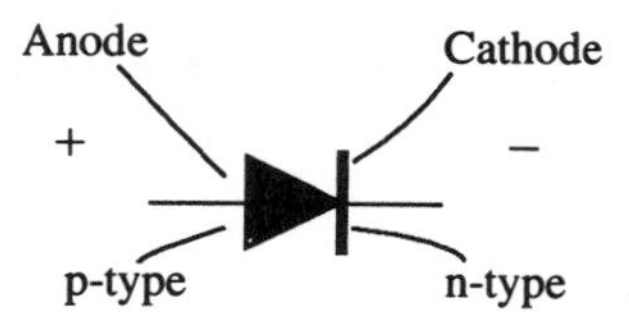

Figure 7.3 *The labels which are sometimes used to identify the electrodes on a p–n junction diode.*

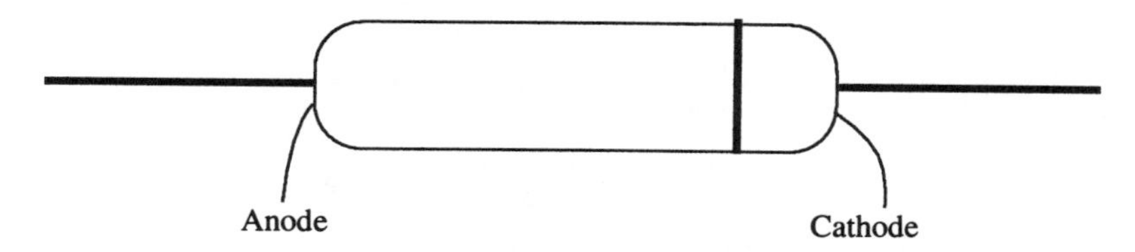

Figure 7.4 *Sketch of a commercial p–n junction diode, showing how a band is sometimes used to identify the cathode.*

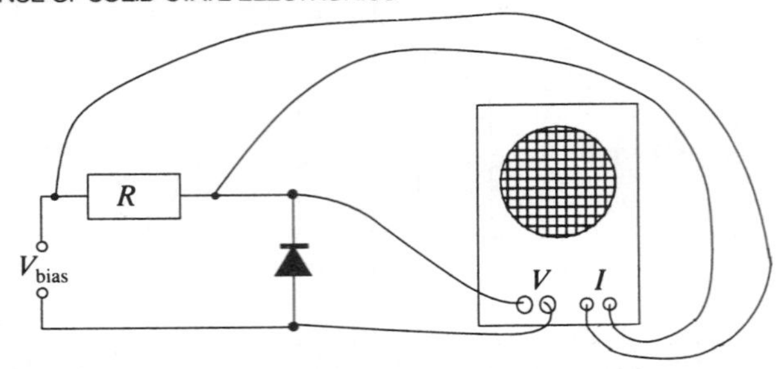

Figure 7.5 *Circuit for observing the I–V characteristic of a p–n junction diode.*

7.1.1 The diode characteristic

Rectification in a p–n junction is seen easily by biasing a p–n junction diode and observing the resulting current–voltage characteristic on a curve tracer (Figure 7.5). Note the inclusion of the resistor R; this is a current-limiting resistor, included to prevent too much current flowing through the diode and damaging it. A current-limiting resistor should always be part of a diode circuit to make sure the diode doesn't overheat — overheating can cause the solder holding the metal leads to the device to melt, thereby causing the leads to fall off (in extreme cases the semiconductor itself could be damaged).

The rectifying nature is seen in Figure 7.6, which is a sketch of a typical p–n junction diode I–V characteristic.

The forward current is much larger than the reverse current because of rectification at the p–n junction. Very often when you observe the current through the diode on an oscilloscope or curve tracer it will appear that the reverse current is zero. This is because the forward current is very large by comparison, and if the oscilloscope is set up to measure forward current it won't be sensitive enough to measure the reverse current. The reverse current is fairly constant with varying reverse voltage whereas the forward current changes rapidly with varying forward voltage. The reverse current has several names: reverse saturation current, reverse leakage current being two of the most common. It's usually given the symbol I_0.

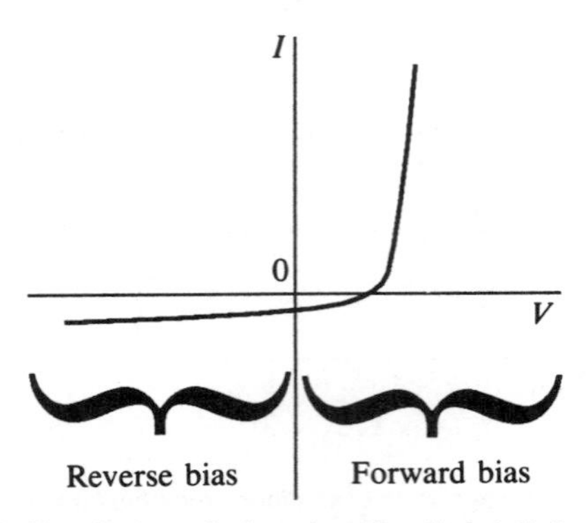

Figure 7.6 *Sketch of a typical p–n junction diode I–V characteristic.*

SELF-ASSESSMENT QUESTION 7.1

Why is the reverse current so small and constant?

7.1.2 The ideal-diode equation

The behaviour of the current through a p–n junction under forward and reverse bias is described by the 'ideal-diode' equation (7.1):

$$I = I_o\{\exp(eV/kT) - 1\} \tag{7.1}$$

where I is the current through the diode, V is the voltage across the diode, I_o is the reverse saturation current, k is Boltzmann's constant, T is the absolute temperature of the diode and e is the electronic charge. Varying the voltage V across a diode and plotting the resulting current I versus V will give a plot similar to that shown in Figure 7.6. Equation (7.1) can be derived from theory (this will not be shown here), but experimental observations have shown that several modifications could be made to equation (7.1) to obtain a closer fit with observations. The 'ideal diode' referred to in the name of the equation doesn't actually exist. An ideal diode is one which has zero forward resistance R_{forward} and infinite reverse resistance R_{reverse} (Figure 7.7). Ohm's law is used to determine the size of the diode resistance under forward bias and under reverse bias, so for an ideal diode zero forward resistance means a vertical forward characteristic and infinite reverse resistance means a horizontal reverse characteristic (Figure 7.7(a)). The ideal diode can be represented as a switch: in forward bias the switch is on (infinite current flows), in reverse bias it is off (zero current flows).

Another model of the ideal diode includes the turn-on voltage $V_{\text{turn-on}}$ (also called threshold voltage) (Figure 7.7(b)). The turn-on voltage is the value of the applied voltage at which the diode starts to conduct in forward bias. The turn-on voltage

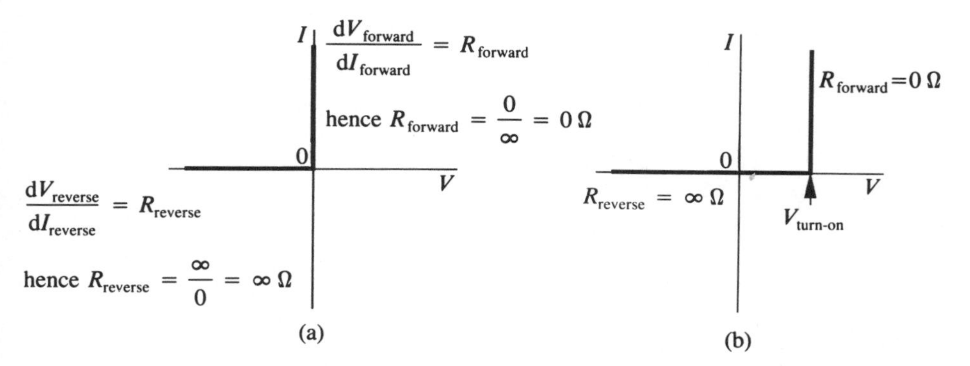

Figure 7.7 The I–V characteristic of an ideal diode: (a) zero forward resistance, infinite reverse resistance and zero turn-on voltage; (b) zero forward resistance, infinite reverse resistance and non-zero turn-on voltage.

depends on the diode material. For example, silicon p–n diodes turn on at about 0.7 V, whereas germanium p–n diodes turn on at about 0.4 V.

SELF-ASSESSMENT QUESTION 7.2
Why do you suppose there is a turn-on voltage?

7.1.3 *Diode applications*

There are a few very important applciations for p–n diodes which depend on how the diode is biased. Figure 7.8 illustrates the different quadrants of the *I*–V characteristic that are used by six important diode types: the rectifying diode, the light-emitting diode, the laser diode, the zener diode, the photodetector, and the solar cell. The conventional quadrant numbering is shown in bold type, and the operating region of each type of diode is shown as a dot on the characteristic curve. Note that the rectifying diode has two dots (one in the first quadrant and one in the third quadrant), whereas the others have one. Quadrants 1 and 4 represent forward bias and quadrant 3 represents reverse bias. Each of the diode types mentioned in Figure 7.8 is described in this chapter.

7.2 Rectifying diodes

The switching behaviour of a diode is used extensively in rectifying circuits, in which the current is allowed to pass through one part of the circuit and not through another, or

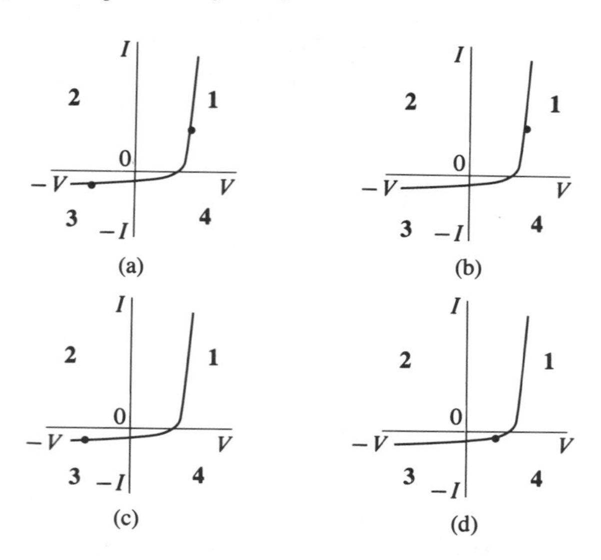

Figure 7.8 The operating quadrants of important diode types: (a) the rectifying diode; (b) the light-emitting diode and laser diode; (c) the zener diode and photodetector; (d) the solar cell.

in a particular direction only. The diodes used in these circuits are commonly made of silicon and have a turn-on voltage of about 0.7 V. For a diode to be a good rectifier, it should have a low value of $R_{forward}$ and a high value of $R_{reverse}$. Rectifying diodes are designed to work effectively under both forward and reverse bias.

A particularly important use for rectifying diodes is in DC power supplies.

7.2.1 ***A simple rectifying circuit***

Rectifying circuits are used to convert alternating current (a.c.) to direct current (d.c.). These circuits are extremely common because most electrical equipment operates on a direct current. For example, a small mains-powered radio will have a plug which connects into the mains socket which supplies electricity at 240 V a.c. and 50 Hz (in the United Kingdom). The radio itself will require a d.c. voltage of a few volts. The radio could use a small battery to obtain the necessary power but batteries are much more expensive than the mains supply. So to avoid unnecessary expense and the inconvenience of having to replace old batteries, the radio will contain a step-down transformer to reduce the size of the voltage and a rectifying circuit to flatten the time-varying a.c. signal, thereby making it d.c.

A very simple rectifying circuit is shown in Figure 7.9. It is a half-wave rectifier or 'clipper', so-called because it removes half of the sine wave coming from the input signal v_{in}. The circuit components are an a.c. supply, a rectifying diode with its current-limiting resistor R, and a load resistor R_L. The load resistor represents the resistance of any circuitry across the circuit terminals.

Consider the sine wave v_{in} coming from the a.c. supply. To find out how the diode rectifies this signal start off by considering only the positive half-cycles of the input signal (Figure 7.10). A positive half-cycle makes the anode (i.e. the p-side) of the diode positive. The diode is thus forward biased and will allow a forward current to flow. So the diode conducts positive half-cycles.

Now consider the negative half-cycles of the input sine wave (Figure 7.11). The anode of the diode is now made negative, hence the diode is reverse-biased. Only a negligible amount of current (I_o) will flow through the diode and we say the diode is turned off: the negative half-cycles are not conducted through the diode. This removal

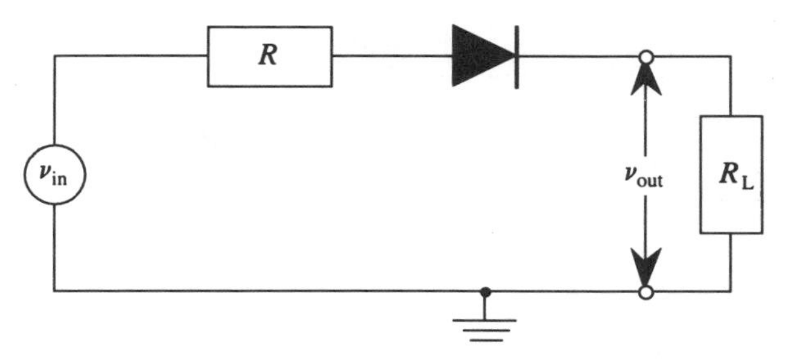

Figure 7.9 *A simple half-wave rectifier circuit.*

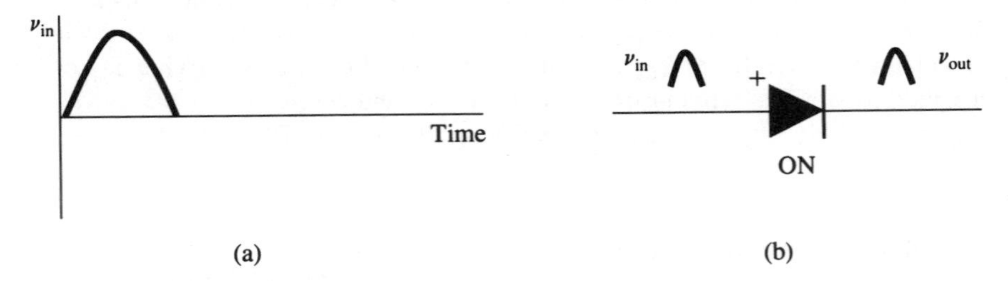

Figure 7.10. *Conduction of positive half-cycles through a p–n diode: (a) the positive half-cycle of the input waveform; (b) the positive half-cycle forward biases the diode.*

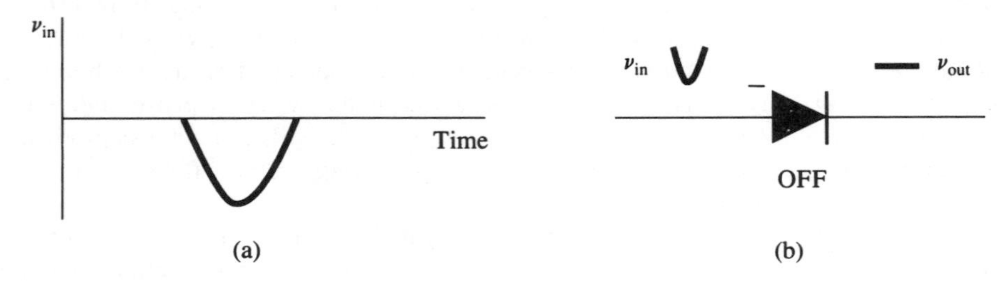

Figure 7.11 *Removal of negative half-cycles by a p–n diode: (a) the negative half-cycle of the input waveform; (b) the negative half-cycle reverse biases the diode.*

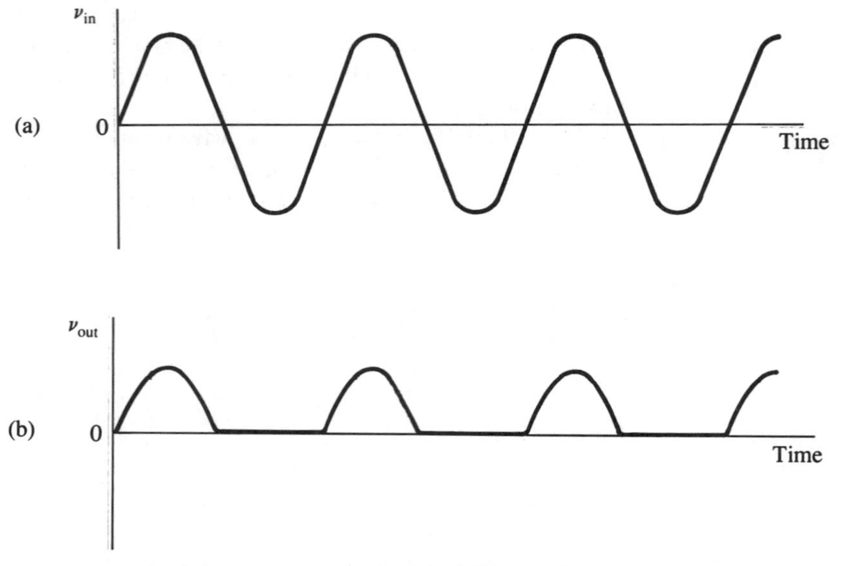

Figure 7.12 *(a) Input and (b) output waveforms of the half-wave rectifier.*

of the negative half-cycles of an input waveform is called clipping. Figure 7.11 shows the input waveform v_{in} and the clipped output waveform v_{out}.

The rectifying circuit just described is very simple. Other types of rectifying circuit, such as the full-wave rectifier, produce an output voltage which is very nearly flat, i.e. almost d.c.

7.3 Zener diodes

So far the illustrations of *I*–V characteristics in this book haven't shown the whole picture. If we were to measure the current through a p–n diode for increasing values of reverse voltage, a point would come where the diode would suddenly start to conduct a large current. So far I've implied that a large current can flow only under forward-bias conditions. This, however, is not the case. At a certain value of reverse voltage all diodes will start to conduct a large reverse current (Figure 7.13). The voltage at which this occurs is called the breakdown voltage V_{br}. The value of the breakdown voltage, like turn-on voltage, depends on the diode material. Some diodes are designed to utilize their breakdown characteristics; they are called zener diodes. In these diodes the reverse breakdown current will increase rapidly over a very small range of reverse voltage. They are therefore used in reverse bias in voltage-regulator circuits and voltage-reference circuits because of their ability to maintain a nearly constant voltage over a wide range of reverse current. The breakdown voltage V_{br} is often referred to as the zener voltage V_Z.

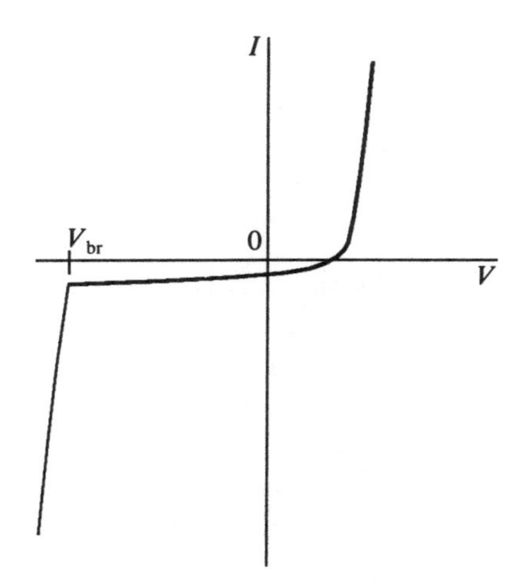

Figure 7.13 *A typical diode I–V characteristic, showing breakdown at the breakdown voltage V_{br}.*

7.3.1 Reverse breakdown

The process which causes the large reverse current to flow is called breakdown. There are two different processes which can lead to breakdown: zener breakdown and avalanche breakdown. Note that diodes which are designed to operate at the reverse breakdown voltage V_{br} are frequently called zener diodes, even if the process they use is the avalanche process.

Zener breakdown

To understand the nature of zener breakdown we need to re-consider the simple model of the p–n junction developed in Chapter 5. Under increasing reverse bias the barrier height between the p-side and the n-side of a p–n junction increases. If the semiconductor is heavily doped such that the p-side Fermi level E_{fp} is extremely low in the bandgap and the n-side Fermi level E_{fn} is extremely high in the bandgap, at low reverse voltages the p-side valence band will lie opposite the n-side conduction band (Figure 7.14). This means the valence electrons on the p-side, of which there will be very many, will be able to cross into the very many empty states on the n-side, *providing the depletion width is narrow enough.* This phenomenon, whereby electrons 'tunnel' through the depletion region from p-side to n-side, is called the zener effect, or tunnelling. A consequence of this is that a large reverse current will occur at relatively low breakdown voltages. A typical value of V_{br} for a zener diode is 5.6 V.

Avalanche breakdown

Reverse breakdown by avalanche occurs at higher reverse voltages than zener breakdown. At a high reverse voltage the depletion width will be large and the built-in electric field $\mathcal{E}$ will be large. A drifting electron from the p-side entering the depletion

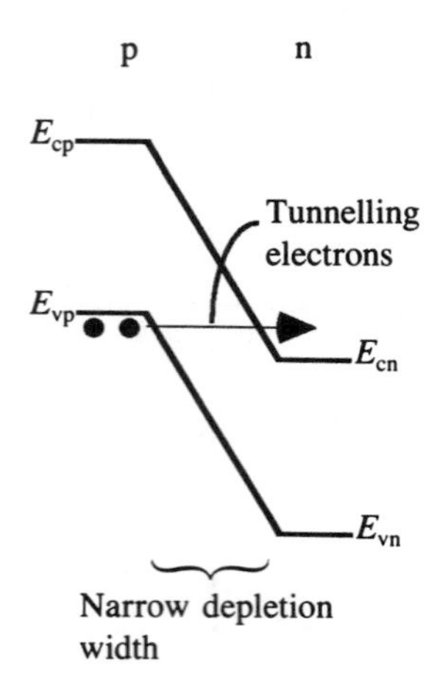

Figure 7.14 *Zener breakdown.*

region will be accelerated through the depletion region because of the electrical force it experiences. The acceleration of the electron could be large enough for the electron to have enough kinetic energy to collide with the lattice and knock another electron out of the lattice, i.e. out of its covalent bond. This second electron will also experience the high electric field, will accelerate and will knock yet another electron out of the lattice (Figure 7.15). Hence there is an avalanche effect as the number of accelerating electrons multiplies rapidly. These electrons which are knocked out of their bonds will be free to take part in conduction processes in the conduction band of the material. Consequently there will quickly be a large number of conduction-band electrons which will form a large reverse current. A typical breakdown voltage for an avalanche diode is 21 V. Large built-in electric fields are required to produce avalanche breakdown, typically of the order of 10^4 V m^{-1}.

EXAMPLE 7.1

Suppose a built-in electric field $\mathcal{E}$ of 5000 V m^{-1} exists across the depletion region of a p–n junction. What acceleration would an electron have on entering the depletion region?

The force experienced by the electron would be

$$\begin{aligned} F_{\text{elec}} &= e\mathcal{E} \\ &= 1.6 \times 10^{-19} \times 5000 \\ &= 8 \times 10^{-16}\ \text{N} \end{aligned}$$

Acceleration is given by Newton's second law of motion:

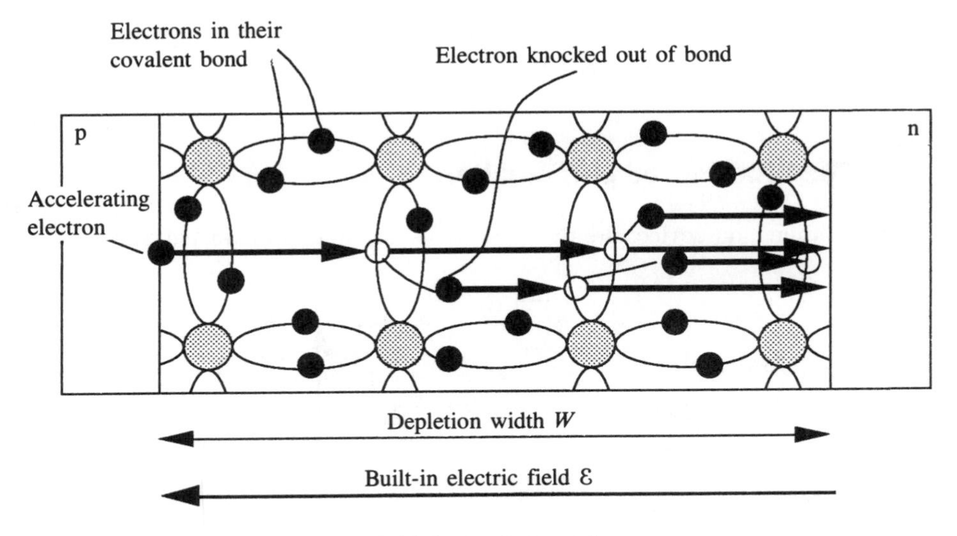

Figure 7.15 *Avalanche breakdown.*

Hence $$\text{Force} = \text{mass} \times \text{acceleration}$$

$$\begin{aligned}\text{acceleration} &= F_{\text{elec}}/m_o \\ &= 8 \times 10^{-16}/9 \times 10^{-31} \\ &= 9 \times 10^{14}\ \text{m s}^{-2}.\end{aligned}$$

The acceleration of the electron will be 9×10^{14} m s^{-2}.

7.3.2 A simple voltage-regulating circuit

For a circuit designer, the important property of a zener or avalanche diode is the suddenness with which the large reverse current flows. This means that, at V_Z, changing the reverse current by a large amount keeps the voltage nearly constant across the diode. Therefore if a reverse biased zener diode is placed across two points in a circuit, those two points will have a nearly constant voltage V_Z dropped between them even though the current may vary. This ability of a zener diode to control or 'regulate' a voltage is very useful. Figure 7.16 shows a simple circuit which can be used to regulate a voltage. Note the circuit symbol for the zener diode.

The voltage regulator of Figure 7.16 consists of a d.c. power supply V_{supply}, a current-limiting resistor R, a load resistor R_L and a zener diode. The zener diode will have a particular value of breakdown voltage V_Z associated with it. The voltage dropped between the two diode terminals will be V_Z providing the zener diode is used within its operating limits. The resistor R should be chosen to make sure the diode is being used in its operating region. You can see in Figure 7.16 that the zener diode is reverse biased. This is essential if the diode is to be used as a voltage regulator. The voltage dropped across the load resistor R_L is fixed by the value of V_Z to give a constant load resistance (called load regulation).

7.4 The illuminated p–n junction

Some p–n junction devices are designed to operate when illuminated, such as the

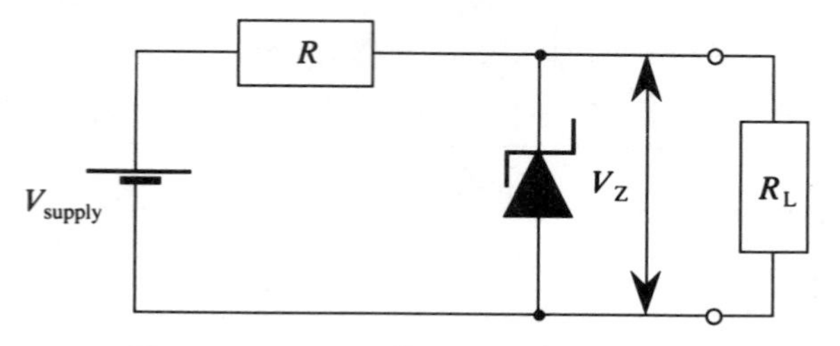

Figure 7.16 *A simple voltage-regulator circuit.*

photodetector (or photodiode) and the solar cell. They are both quite different in the way they are used because the photodetector is designed to detect light whereas the solar cell is designed to generate power. However, the principle of operation is the same in each case. Consider a p–n junction which is shielded from the light. This junction will behave like the p–n junction described in Chapter 5: under zero-bias conditions the diffusion current (majority carriers) across the depletion region is counterbalanced by a drift current (minority carriers) in the opposite direction. Under forward bias the depletion region narrows, the number of majority carriers crossing the energy barrier increases, and there is an increase in the forward diffusion current; the drift current is unchanged. Under reverse bias the depletion region widens and the number of majority carriers crossing the energy barrier decreases such that the diffusion current becomes negligible; the drift current is unchanged.

When the p–n junction is illuminated majority and minority carriers are optically generated in the n-side and in the p-side. The minority carriers so created are swept across the depletion region by drift, providing they are generated within a diffusion length of the depletion region. The creation of drifting minority carriers using this method is called optical *injection*. A large reverse saturation current can be produced by this means. The minority-hole current from n to p would then depend on the rate of hole injection into the n-side within L_p of the depletion region, and the minority-electron current from p to n would then depend on the rate of electron injection into the p-side within L_n of the depletion region.

These optically generated carriers are called 'excess' carriers because they are extra to those generated thermally. The creation of excess carriers means that there will be more electrons in the conduction band and more holes in the valence band of the p–n junction, so increasing the numbers of majority and minority carriers. Under forward bias the diffusion and drift currents will increase, the drift current being swamped by the diffusion, as usual. Under reverse bias the amount of diffusion remains negligible, but the minority drift current increases dramatically. The behaviour of the minority carriers in the illuminated p–n junction is particularly important because of this. Figure 7.17 is a sketch of the I–V characteristic of a p–n junction under dark and light conditions, showing how increasing illumination levels increase the reverse saturation current. Notice the different scales for the forward and the reverse currents: I is graduated in milliamperes but $-I$ is graduated in microamperes. The dark characteristic is formed by thermally generated minority carriers drifting across the depletion region. The characteristics under increasing illumination are formed by thermally generated *and* optically generated minority carriers drifting across the depletion region.

The I–V characteristic changes significantly under reverse bias as the intensity of illumination is increased. The reverse current, called the photocurrent, may increase a thousandfold.

Because p–n junctions react to light as described here it's essential that rectifying and zener diodes are shielded from the light, hence they're surrounded by a resin which is electrically inert and then by an opaque plastic case. Consequently you never see the actual diode itself — you see only the package.

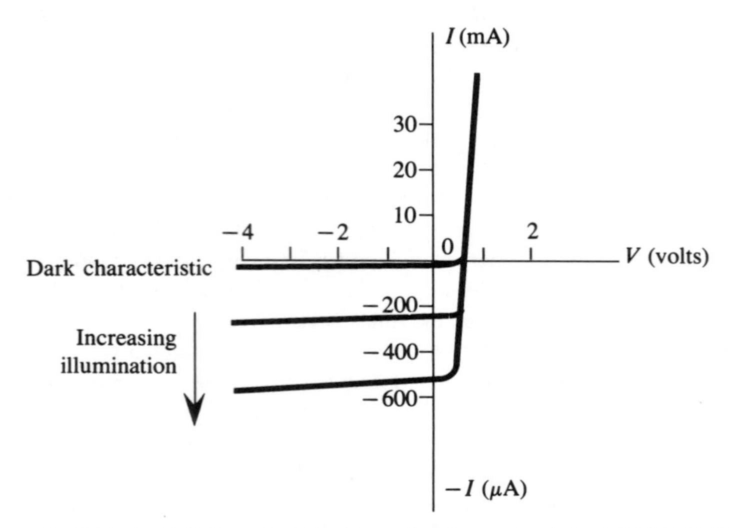

Figure 7.17 *Typical I–V characteristic of a p–n junction under dark and light conditions.*

It is the increasing reverse current which is the basis for the two diodes we're going to consider next — the photodetector and the solar cell.

7.4.1 Photodetectors

Photodetectors are light detectors. They are operated in reverse bias and utilize the increasing reverse current of the I–V characteristic. In circuit terms photodetectors are thought of as light-dependent resistors because the resistance R_{reverse} decreases as the light intensity hitting the junction increases.

SELF-ASSESSMENT QUESTION 7.3

Using Figure 7.17, what is R_{reverse} under dark conditions? What is the minimum R_{reverse} under light conditions?

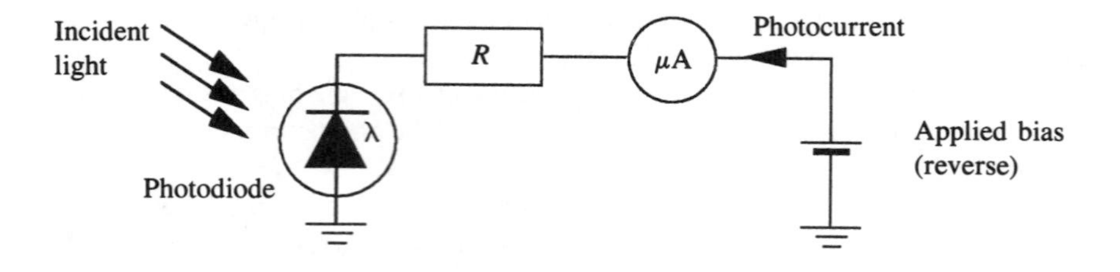

Figure 7.18 *A simple photodetector circuit.*

Figure 7.18 shows how a photodetector would be connected to a microammeter and a battery or power supply to be used as a light detector. As the light intensity is increased a larger reverse current would be measured on the microammeter. Note the current-limiting resistor R, included to protect the photodetector.

Photodetectors are particularly interesting because of their structure. Unlike rectifying diodes, it's essential that light gets into the junction so the semiconductor is housed in a mounting which has a transparent window; this means you can see the surface of the semiconductor itself, and if you look carefully you should even be able to see the tiny wires which connect the semiconductor to the outside world. Another structural difference between the photodetector and the rectifying diode is the size of the junction region itself. In a photodetector the junction needs as large a surface area as possible to maximise the number of photons hitting the surface. Also, the junction needs to be close to the surface in the photodetector. If the carriers are to recombine in the junction region they must be able to travel that far, so the distance of the junction from the surface must be less than the diffusion length. Lastly, a common design requirement is the deposition of an anti-reflection coating on the semiconductor surface. This increases the number of photons which are absorbed by the diode.

The p–i–n diode is one of the most common types of photodetector (Figure 7.19). It consists of a p–n junction separated by a wide depletion layer, hence the 'i' meaning 'intrinsic' or 'insulating' in the name. The depletion region is wide so that a large electric field $\mathcal{E}$ acts across it. This field $\mathcal{E}$ then sweeps the excess holes and electrons across the depletion width by drift so that they can contribute to the photocurrent. Other features shown in Figure 7.19 are the two metal electrodes, one to the p-side and one to the n-side, and the silicon dioxide patches which are used in integrated circuits to insulate the surface of one device from its neighbours. Silicon dioxide is an excellent insulator because of its wide bandgap (Table 3.1), so it is used as an insulator in electronic devices.

The most important design consideration, however, is that the diode detects over a desired wavelength range with enough sensitivity to enable practical use. The long-wavelength cut-off point depends on the bandgap of the semiconductor.

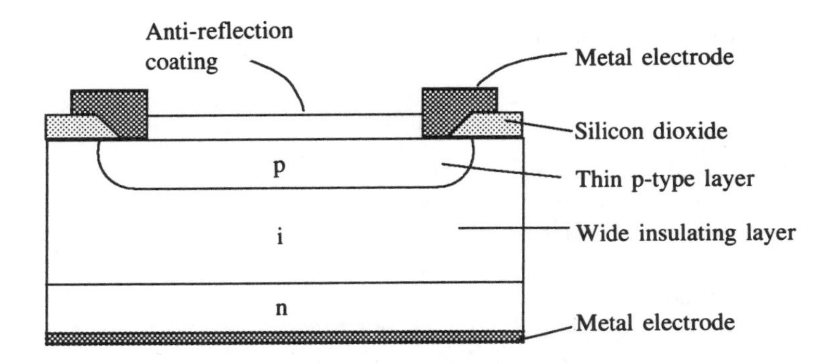

Figure 7.19 *Basic p–i–n photodetector structure.*

EXAMPLE 7.2

If you wish to make a photodetector to detect infra-red, say from 1 μm upwards, the semiconductor must be capable of absorbing photons of wavelength 1 μm or more. For such a photodetector, then,

$$E_{ph(max)} = \frac{hc}{e\lambda_{min}} \text{ eV} \tag{7.2}$$

where λ_{min} is 1 μm. The bandgap E_g must therefore be given by

$$Eg \leq E_{ph(max)}$$

Hence

$$E_g \leq \frac{hc}{e\lambda_{min}}$$

$$\leq \frac{6.6 \times 10^{-34} \times 3 \times 10^{8}}{1.6 \times 10^{-19} \times 1 \times 10^{-6}} = 1.2 \text{ eV}$$

Hence the bandgap required to detect photons of at least 1 μm is less than or equal to 1.2 eV. Silicon might therefore be a suitable semiconductor for such a photodetector.

p–i–n detectors made of gallium indium arsenide GaInAs and germanium Ge are commonly used to detect the 1.3 μm and 1.55 μm infra-red used in optical-fibre communication systems.

SELF-ASSESSMENT QUESTION 7.4

If the bandgap of silicon is 1.1 eV, what is its long-wavelength cut-off? Give your answer in nanometres.

7.4.2 **Solar cells**

Solar cells are used to generate power. Electrical power is given by the simple expression

$$\text{power} = I\ V$$

Hence, if V is positive but I is negative the power will be negative. A negative power means that power is generated, not consumed. This means that a diode can be used to deliver power to an external circuit providing it is forward biased *with a reverse current flowing*. There is one quadrant, the fourth, of the I–V characteristic where this is possible (Figure 7.20).

Solar cells are designed to generate as much power as possible when the cell is illuminated. As with the photodetector, there are several important design considerations to note: the bandgap must allow absorption of suitable

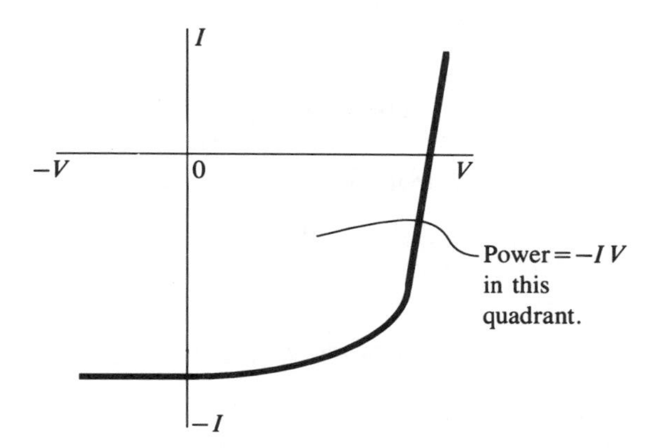

Figure 7.20 *The fourth, or solar-cell, quadrant of the diode I–V characteristic.*

wavelengths (sunlight produces a lot of infra-red, for instance); the surface area must be large to allow as many photons to be absorbed as possible; the junction must be close to the surface so that diffusing carriers reach the junction before recombining. Figure 7.21 shows the structure of a simple solar cell. Any surface electrodes must have a very small area compared to the surface area exposed to light to maximize the number of incident photons. At the same time the area covered by the electrode needs to be as large as possible to reduce detrimental resistance effects. These two contradictory requirements are met by a compromise solution of giving the top electrode a comb-like structure (Figure 7.21(b)). Also, the semiconductor surface should be coated with an anti-reflection coating to increase the number of photons being absorbed rather than reflected. Individual solar cells are not capable of delivering enough power for most purposes, so they are commonly used in the form of large arrays of cells connected together.

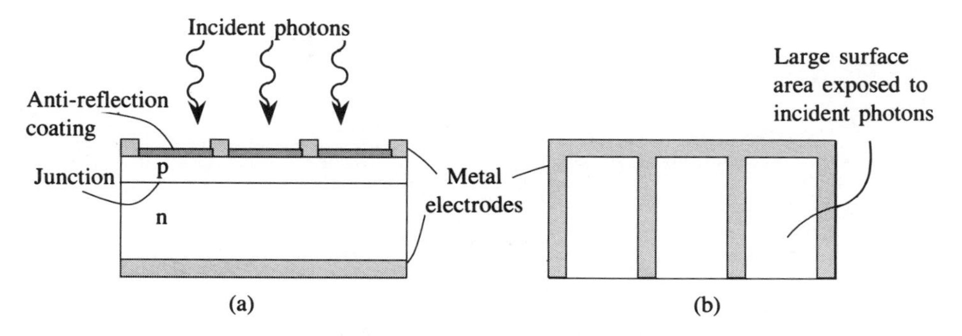

Figure 7.21 *Simple solar cell structure: (a) side view; (b) top view.*

Fill factor

Many optoelectronic devices have a figure of merit associated with them which is used to indicate how good the device is. In the case of the solar cell, the figure of merit is the fill factor:

$$\text{Fill factor} = \frac{P_{max}}{I_{sc}V_{oc}}$$

where P_{max} is the maximum output power from the cell, given by

$$P_{max} = I_{max}V_{max}$$

Hence

$$\text{Fill factor} = \frac{I_{max}V_{max}}{I_{sc}V_{oc}} \tag{7.3}$$

where I_{sc} is the short-circuit current (the value of I when V is zero) and V_{oc} is the open-circuit voltage (the value of V when I is zero). The quantity P_{max} is known as the maximum-power rectangle and is shown in Figure 7.22. P_{max} will always be less than the product $I_{sc}V_{oc}$, so the fill factor will always lie between zero and one.

SELF-ASSESSMENT QUESTION 7.5

What is the fill factor for the solar cell characteristic shown in Figure 7.23?

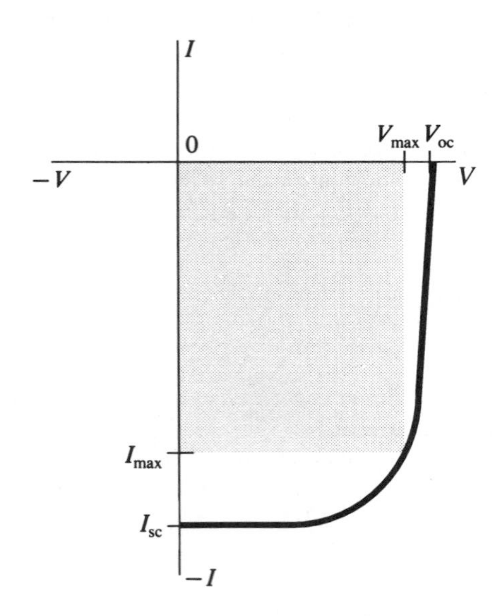

Figure 7.22 *The maximum-power rectangle $V_{max}I_{max}$ for an illuminated solar cell.*

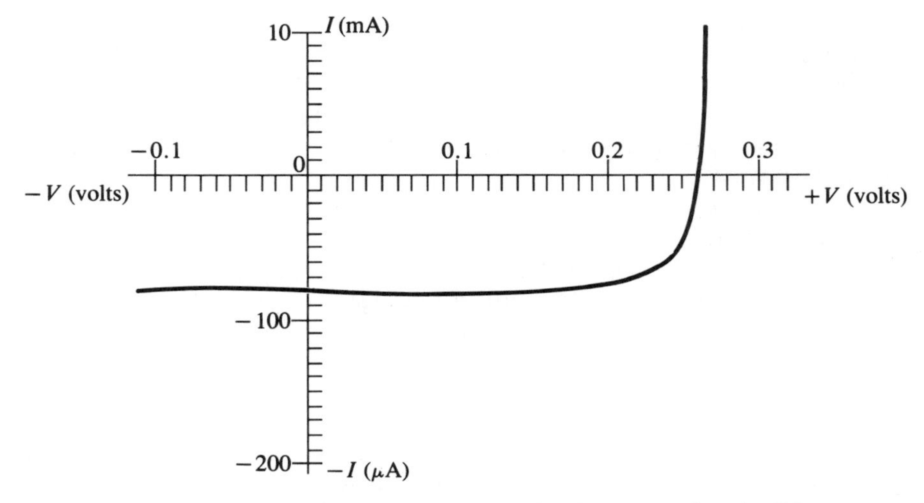

Figure 7.23 *The solar-cell characteristic for Self Assessment Question 7.5.*

Efficiency

The efficiency of a solar cell is often referred to as solar efficiency, η_s, given by the expression

$$\eta_s = P_{max}/P_{solar} \tag{7.4}$$

where P_{solar} is the amount of solar power reaching the cell. Ideally, η_s would be equal to one, indicating that all incident solar radiation is being converted to electrical power by the cell.

Despite well-known design requirements the efficiency of solar cells remains low, at less than 30%. Gallium arsenide is a favoured material for solar cells because of its high efficiency.

7.5 Light-emitting junction devices

Light-emitting semiconductor devices such as the light-emitting diode (LED) and the laser diode utilize the phenomenon of recombination to produce the light, a process called injection electroluminescence. In a forward-biased p–n junction there is a large diffusion current across the junction region. This diffusion current consists of holes moving one way and electrons moving in the opposite direction, so that considerable recombination will occur. In a semiconductor with an indirect bandgap, such as silicon or germanium, the recombination of EHPs in the junction region produces heat which is dissipated in the lattice. In a direct-bandgap semiconductor, however, such as gallium arsenide or gallium indium arsenide phosphide, the recombination is very efficient and produces photons which have an energy equal in size to the transition energy (refer back to Chapter 6 to remind yourself about direct and indirect semiconductors).

For example, in gallium arsenide a bandgap transition means that the recombination will be across the bandgap — an electron falls from the conduction band to the valence band, a transition of 1.4 eV at room temperature. In the same material, if it is n-type, recombination could also take place between the conduction band and a donor site. This transition energy will be much smaller, typically 0.006 eV.

SELF-ASSESSMENT QUESTION 7.6
What wavelengths would be obtained for transitions in gallium arsenide? Assume recombination occurs across the bandgap (1.4 eV) and from the conduction band to a donor site (assume 0.006 eV below the conduction-band edge).

Light-emitting p–n junction devices can operate efficiently at room temperature and can be made very small, typically less than 1 mm long, which makes them ideal light sources for displays or optical communications.

7.5.1 Light-emitting diodes

Displays using LEDs are very commonly seen in pocket calculators, at airports and railway stations, etc. Of course, making a display board means it's essential to have LEDs which emit in the visible spectrum. It's now possible to fabricate LEDs with wavelengths extending right through the visible spectrum from red to violet, by using compound semiconductors and alloying them together in different proportions. Red LEDs are commonly made of gallium arsenide phosphide GaAsP, a compound which is produced by alloying gallium arsenide GaAs and gallium phosphide GaP. The alloy $GaAs_{0.6}P_{0.4}$ is commonly used to produce photons of energy 1.8 eV (equivalent to a wavelength of 0.69 μm).

The basic design of an LED is quite simple. Preferably it requires a direct-bandgap material, a p–n junction which can be forward biased via two electrodes, and a way of getting the emitted photons out of the device without being absorbed by the material. Figure 7.24 is a sketch of a simple LED. When this LED is forward biased a large number of diffusing electrons will be injected from the n-side into the p^+-side where the density of holes is very high. Considerable recombination is obtained in this way, and photons will be emitted from the junction region. The n^+ layer in the LED shown is there to give a good electrical contact with the bottom electrode.

7.5.2 Laser diodes

The emission in the LED is called spontaneous because there is no control over the recombination of electrons and holes. Electrons are excited into the conduction band at any time, they recombine spontaneously, and there needn't be many of them recombining at any one time. If we were to control the carriers such that there were more electrons waiting to recombine than electrons waiting to be excited at any one

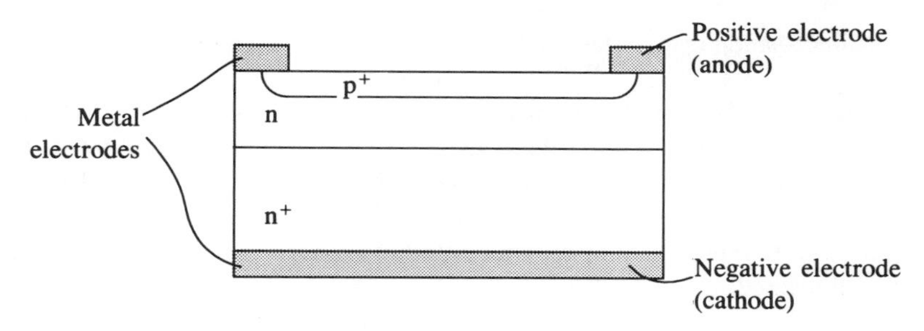

Figure 7.24 *Simple LED structure.*

time (a phenomenon called population inversion), we could then arrange for the recombination of all these simultaneously. This would produce a more intense beam of emitted light. This is what happens in the laser diode. The name is derived from light amplification by the stimulated emission of radiation.

A laser diode is similar to an LED in that it is essentially a forward-biased p–n junction in which recombining carriers produce light. There are important differences, however. In the laser the excited electrons are *stimulated* to recombine simultaneously and produce an intense beam of photons of the same wavelength. Some of these photons are then reflected back through the device to generate more electron–hole pairs, which in turn recombine to produce more light, which is reflected back through the device, and so on. Hence in a laser the reflection of photons back and forth through the device produces amplification of the emitted light. Reflection is achieved by polishing the end faces of the laser.This is quite straightforward, even in such small devices; when a semiconductor is cleaved along the (100) plane, for example, the newly exposed surfaces are very shiny and mirror-like. The ends may then be chemically treated or coated with metal to retain their mirror finish. Figure 7.25 is a schematic diagram of a simple laser diode, showing the mirrored end faces and the reflection and emission of photons. The depletion region is called the active region in a laser diode. As with the LED, direct transitions are much more efficient than indirect

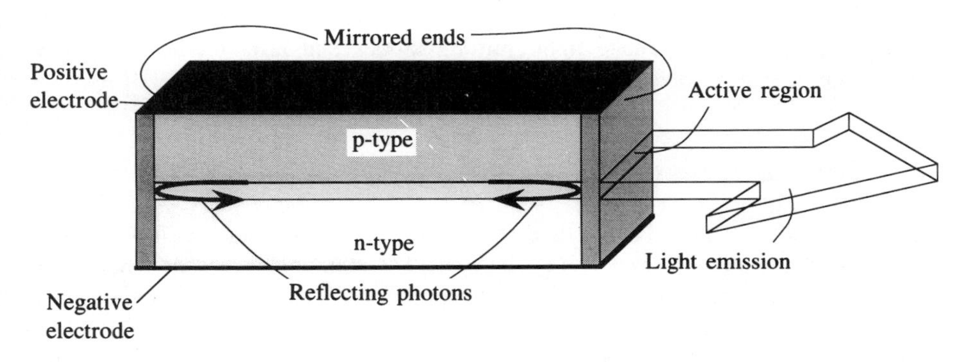

Figure 7.25 *Simple laser-diode structure.*

ones so direct-bandgap materials such as gallium indium arsenide phosphide GaInAsP are used.

As a result of the stimulated emission and the amplification of the light, a laser diode emits a more intense beam than the LED. Also, the light emitted by the laser is much more monochromatic than that produced by the LED. Laser diodes do have disadvantages: they are more expensive than LEDs, they are more complex to design, more difficult to fabricate, they are more prone to failure, and their drive circuitry is more complicated.

7.6 Summary

The I–V characteristic describes the electronic behaviour of the p–n diode, and various electronic devices are made which utilize the properties shown by the characteristic. The diode's switching ability is used by rectifying diodes which are made to operate in both forward and reverse bias, whereby the negative half-cycles of an a.c. input signal are removed at the output (called clipping). Zener diodes are designed to operate at the reverse-breakdown voltage (or zener voltage). While in reverse bias a zener diode maintains a constant voltage across its terminals even though the current through the diode may change. This phenomenon is used in voltage regulators. There are two reverse-breakdown processes: zener breakdown and avalanche breakdown.

When a p–n diode is illuminated the reverse saturation current increases. As the intensity of illumination increases, so does the reverse current. Photodetectors use this property as a way of detecting light. The p–i–n diode is an example of such a photodetector, in which optically generated excess carriers form the increased reverse current. When a p–n diode is illuminated it is possible to operate the diode in the fourth quadrant of the I–V characteristic where the applied voltage is positive but the current is negative, hence the diode (or solar cell) generates power.

When operated under foward bias the p–n diode exhibits injection electroluminescence, whereby photons are electrically produced when carriers of one type are injected across a p–n junction to recombine with carriers of the opposite type. This recombination produces photons which are equal in energy to the transition energy. It's possible to make light-emitting diodes and laser diodes using this phenomenon. Many wavelengths are available, depending on the material used.

7.7 Tutorial questions

7.1 Use equation (7.1) to sketch a diode I–V characteristic. Assume I_o is 1 nA and T is room temperature. Vary V from -10 V to $+1$ V. Remember that the units of eV must be the same as the units of kT: if the product eV is in electron volts, kT must also be in electron volts.

7.2 Consider the ideal-diode equation (7.1). If a forward voltage V_f is applied to the diode, such that

$$V_f \geq 3kT$$

what size will the current I through the diode be? Use equation (7.1) to show that this current would increase exponentially with increasing forward bias. Likewise, if the reverse bias V_r were

$$V_r \geq -3kT$$

what size will the current I through the diode be? Use equation (7.1) to show that this current would remain approximately constant with increasing reverse bias.

7.3 Explain why gallium arsenide GaAs is a more suitable material than silicon Si for the fabrication of optoelectronic devices such as light-emitting diodes and laser diodes.

7.4 An engineer wishes to design a diode that emits red light. Which of the six alloys shown in Table 7.1 could she use, and why? Illustrate your answer by the use of appropriate equations and calculations.

Table 7.1 *Alternative alloys*

Alloy	*Bandgap type*	*Bandgap (eV)*
$Al_{0.4}Ga_{0.6}As$	Indirect	1.96
AlAs	Indirect	2.16
GaP	Indirect	2.26
InP	Direct	1.35
$Al_{0.3}Ga_{0.7}As$	Direct	1.87
$GaAs_{0.6}P_{0.4}$	Direct	1.94

7.5 Sketch the I–V characteristic of an avalanche diode, indicating the breakdown region. How would you expect the I–V characteristic of a zener diode to differ from that of an avalanche diode?

7.6 Sketch the I–V characteristic of a p–n diode and mark on the quadrants which contain the operating regions of the following devices:

- p–i–n diode
- laser diode
- solar cell
- light-emitting diode
- avalanche diode

7.8 Suggested further reading

Boylestad, R. and Nashelsky, L., *Electronic Devices and Circuit Theory*, 5th edn, Prentice Hall, 1992, pp. 24–31, 51–99, 798–827.

Floyd, T. L., *Electronic Devices*, 3rd edn, Merrill, 1992, pp. 37–50, 98–110, 113–118.

Grovenor, C.R.M., *Microelectronic Materials*, Adam Hilger, 1989, pp. 373–388, 393–412, 422–428, 430–442.

Parker, G., *Introductory Semiconductor Device Physics*, Prentice Hall, 1994, pp. 224–235.
Streetman, B. G., *Solid State Electronic Devices*, 4th edn, Prentice Hall, 1995, pp. 212–227, 391–392.
Wilson, J. and Hawkes, J.F.B., *Optoelectronics: an Introduction*, 2nd edn, Prentice Hall, 1989, pp. 131–137, 155–157.

CHAPTER 8

The field-effect transistor and its operation

Aims and objectives

This chapter follows the development of p–n junction devices started earlier in the book by examining the operation of junction field-effect transistors and using equations describing their electrical performance to determine commonly used FET parameters such as pinch-off voltage. MOSFET operation is also described.

8.1 Transistors

Transistors are semiconductor devices which have three terminals and which have the ability to amplify an input signal. A voltage is applied between two of the terminals so that a current flows between them, and the third terminal is used to control the size of that current. There are two major types of transistor: the bipolar transistor (also called the bipolar junction transistor or BJT) and the field-effect transistor (or FET or unipolar transistor), which is sometimes called the unipolar transistor because it relies on the transport of just one carrier type for its operation. The FET is a voltage-controlled device and its operation depends on the control of electric fields in the semiconductor material: the output current is controlled by a small input voltage. The BJT, on the other hand, uses an input current for its control and uses both carrier types for its operation. The BJT is described in Chapter 9.

There are several types of FET which could be described here. The simplest to model is the junction field-effect transistor (JFET or junction-gate FET or JUGFET), so that will be described first to illustrate basic FET operation. This will be followed by a brief description of the MOSFET (or IGFET).

All transistors are three-terminal devices, i.e. they have three electrodes. In the FET these three electrodes are always called:

- drain
- source
- gate

8.2 The junction FET

The FET operates by passing a current through a pathway in the device called a channel. This channel is used to classify the JFET, of which there are two types: n-channel and p-channel. Figure 8.1 is a schematic diagram showing cross-sections through an n-channel JFET and a p-channel JFET. You're unlikely to use a FET having a cylindrical structure like the one shown, but the cylinder shape is easy to describe and therefore makes FET operation easier to understand. Figure 8.1 shows the three electrodes that all FETs have: the drain, the gate and the source. These electrodes would typically be made of a metal such as aluminium, copper or gold. The device itself is made of a single semiconductor crystal which contains a p–n junction. The three wire leads are attached to the semiconductor at the three metal electrodes.

In the n-channel JFET, the gate region is made of a p-type material and the rest of the device is made of n-type. The gate region is represented as a narrow p-type 'belt' around the middle of a cylinder. The drain and the source form each end of the cylinder and are connected by a channel in the n-type material. p–n junctions lie between the drain and the gate and between the source and the gate.

In the p-channel JFET, on the other hand, the gate region is made of an n-type semiconductor and the rest of the device is made of p-type. The drain and the source are therefore connected by a p-type channel. There are p–n junctions between the drain and the gate and between the source and the gate.

The most common circuit symbols for the n-channel and the p-channel JFETs are shown in Figure 8.2. Notice the direction of the arrowhead on each gate electrode.

Look again at Figure 8.1. The gate forms a p–n junction with the drain and also with the source. These two p–n junctions in the JFET are responsible for its operation.

Operation of the n-channel JFET

Consider the n-channel JFET illustrated in Figure 8.1(a). In order for a current to flow through the length of the device we need to apply a voltage between the drain and the source, so connect them by an applied bias V_{DS} such that the drain is positive relative to

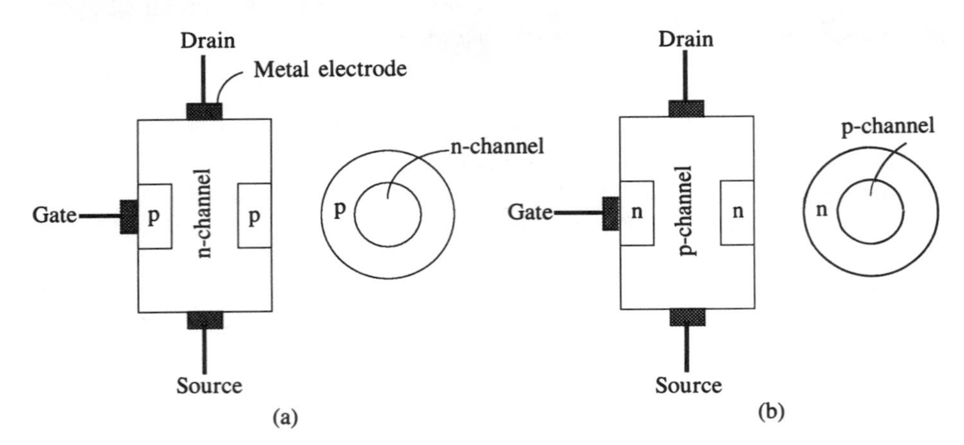

Figure 8.1 *Schematic diagrams showing cross-sections through (a) n-channel and (b) p-channel junction field-effect transistors.*

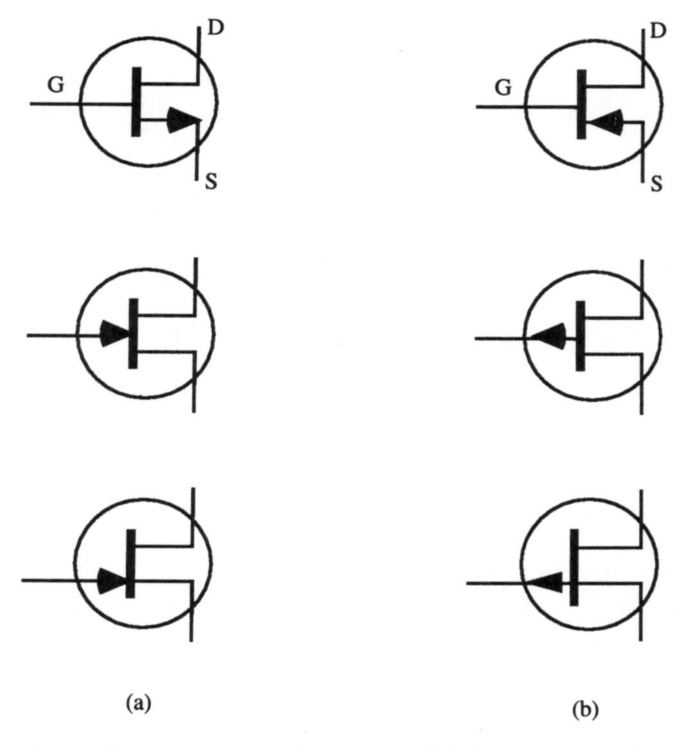

Figure 8.2 *Symbols for (a) the n-channel JFET, (b) the p-channel JFET.*

the source (Figure 8.3). This means that electrons in the n-type channel will flow from the source towards the drain; hence a conventional current flows from the drain to the source. We call this current the drain current and give it the symbol I_D. The size of I_D will depend on the size of V_{DS}, according to Ohm's law:

$$I_D = V_{DS}/R_{channel} \tag{8.1}$$

where $R_{channel}$ is the resistance of the channel between the drain and the source. With V_{DS} connected, therefore, the JFET behaves like a resistor of resistance $R_{channel}$. There is no p–n junction between the drain and the source. Note the polarity of V_{DS}: the current formed in this JFET will be in the direction drain→source. (Remember that current travels from the positive pole of the battery, through the circuit, and so to the negative pole of the battery.) This current is called the drain current I_D (Figure 8.3). If we consider carrier flow instead of current flow, in the n-channel JFET the electrons forming the drain current flow from the source to the drain (Figure 8.3).

SELF-ASSESSMENT QUESTION 8.1

What are the majority carriers in the n-channel?

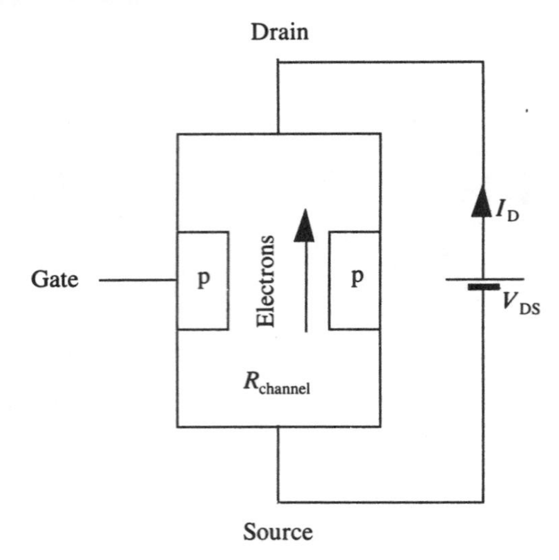

Figure 8.3 *A bias voltage V_{DS} applied between drain and source to cause a current I_D to flow through an n-channel JFET.*

SELF-ASSESSMENT QUESTION 8.2

Figure 8.4 shows a p-channel JFET with applied bias V_{DS}. What are the directions of the current flow (I_D) and the majority-carrier flow? What are the majority carriers in the channel?

So far we've just considered what happens between the drain and the source. What about the gate? If the gate region were removed the JFET would behave just like a resistor — current would flow straight through the device, the amount depending on the resistance of the semiconductor material. So for a given value of V_{DS} we'll be able to control the size of I_D if we can vary $R_{channel}$. We'll find that the gate is used to control the channel resistance and hence control the size of the drain current I_D. This variation in $R_{channel}$ is achieved using the p–n junction between the gate and the source. This junction is always reverse biased in the JFET, so the gate is made negative relative to the source in the n-channel JFET (Figure 8.5). This bias voltage between the gate and the source is known as V_{GS}. It is this bias voltage which is used to control $R_{channel}$.

Remember what applied bias does to a p–n junction: forward bias decreases the depletion width whereas reverse bias increases it. Hence varying gate voltage from zero volts to minus a few volts will enable control over the size of the depletion width around the gate region. The gate has a high dopant concentration whereas the channel is lightly doped.

SELF-ASSESSMENT QUESTION 8.3

Why do you suppose the gate is highly doped and the channel is lightly doped? (You may need to refer back to Chapter 5 to answer this.)

To control the size of the depletion layer

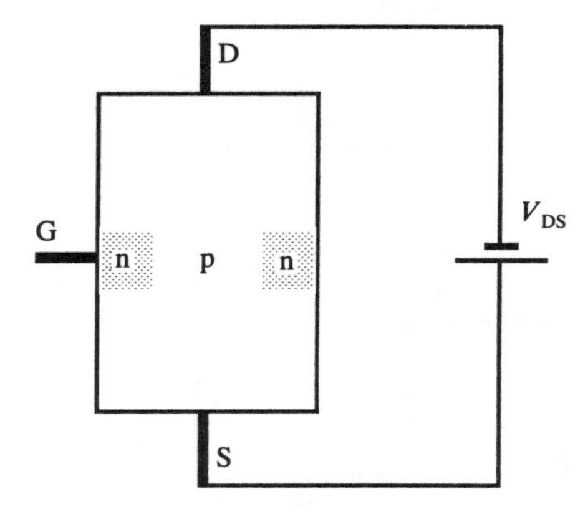

Figure 8.4 *A p-channel JFET with applied bias V_{DS}.*

Before we move on, we should consider our schematic n-channel JFET (Figure 8.6). There are three voltages which can be applied across the three electrodes: V_{DS}, V_{GS} and V_{GD}.

Now consider an n-channel JFET with no applied bias (i.e. $V_{DS}=0$) (Figure 8.7) and remember that in an abrupt p–n junction at equilibrium (i.e. zero bias) there is a depletion region across the junction. There will therefore be a depletion region around the gate in the n-channel JFET. This depletion region will extend outwards from the p-type gate region into the n-channel. The depletion region contains no free carriers so the parts of the channel it occupies behave like an insulator. If this depletion region

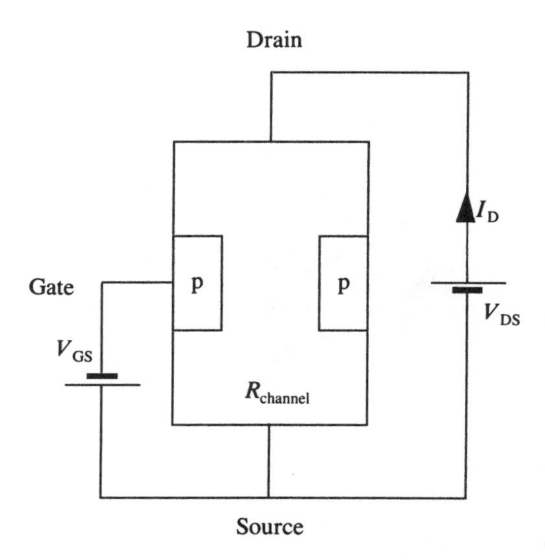

Figure 8.5 *Using an applied bias V_{GS} to reverse bias the gate in the n-channel JFET.*

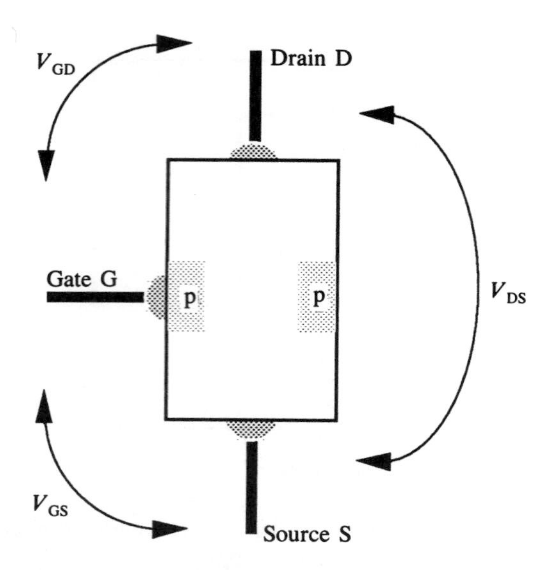

Figure 8.6 *The three voltages associated with a JFET.*

were made to grow larger, the channel resistance would effectively increase. Likewise, if the depletion region shrank, then the channel resistance would effectively decrease. In other words, if the depletion region were larger, less drain current would flow because the channel would be smaller; if the depletion region were smaller then more current would flow because the channel would be wider.

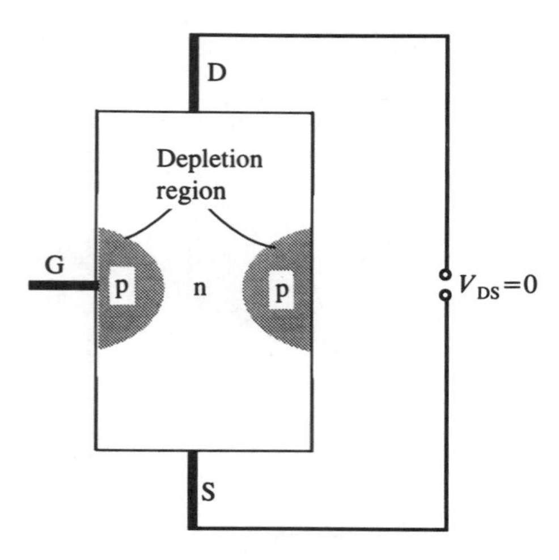

Figure 8.7 *An unbiased n-channel JFET, showing the depletion region around the gate.*

Now make V_{DS} a small voltage (Figure 8.8) such that the drain is positive relative to the source. We know from our previous discussion that a drain current I_D will flow from the drain to the source. The size of this current will depend on the resistance of the channel, or how far the p–n junction around the gate extends into the n-type material. The equilibrium or zero-bias case no longer applies because of V_{DS}; the p–n junction between the gate and the drain is reverse biased and that between the gate and the source is forward biased. This changes the appearance of the depletion region from that indicated in Figure 8.7. The gate–drain junction is reverse biased, therefore the depletion region there is relatively wide. The gate–source junction is forward biased, therefore the depletion region there is narrow. This gives a lopsided appearance to the depletion region around the gate (Figure 8.8).

SELF-ASSESSMENT QUESTION 8.4

Why is the gate–drain junction reverse biased and the gate–source junction forward biased?

So much for V_{DS}, but what about the gate? In fact the gate is used to control the size of the channel (and therefore the channel resistance) by controlling the size of the depletion region. Bias the gate–source junction by applying a voltage V_{GS} between gate and source such that the gate is negative relative to the source (Figure 8.9). Note that V_{GS} and V_{DS} are both measured relative to the source; this can be achieved by grounding the source.

Now suppose the size of V_{GS} is increased from zero, say from 0 V to −1 V. (Note the polarity of this voltage! If V_{DS} is a positive voltage then V_{GS} must be negative.) This means the junction between the gate and the drain is more reverse biased than previously, so that the depletion region between the gate and the drain will be larger.

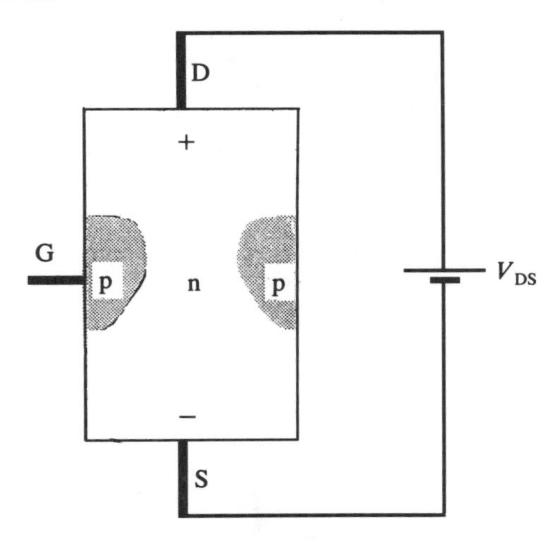

Figure 8.8 The same JFET, showing how the depletion region extends further into the channel towards the drain when V_{DS} is applied.

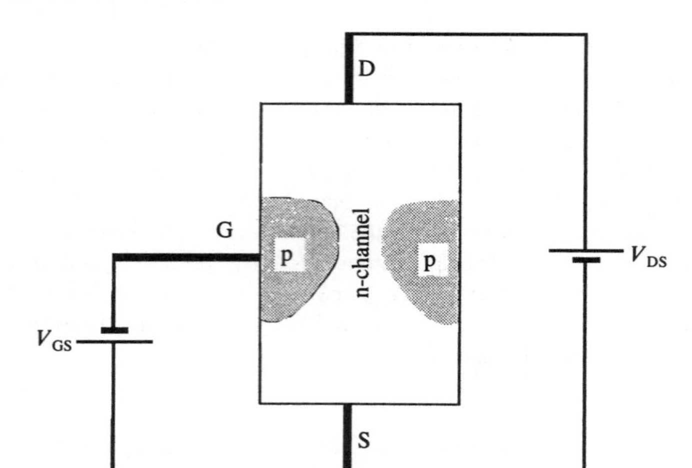

Figure 8.9 *The gate is made negative relative to the source.*

The depletion region between the gate and the source also will be larger. Hence the increased value of V_{GS} has resulted in a larger depletion region around the gate and therefore a smaller channel. This means $R_{channel}$ increases and the drain current I_D decreases.

n-channel JFET drain characteristic

The drain characteristic of a FET is a plot of drain current I_D versus drain–source voltage V_{DS}. The characteristic shows how the drain current I_D varies with V_{DS}, for a range of V_{GS} values. So, to plot the graph consider the variation of I_D with V_{DS} for zero

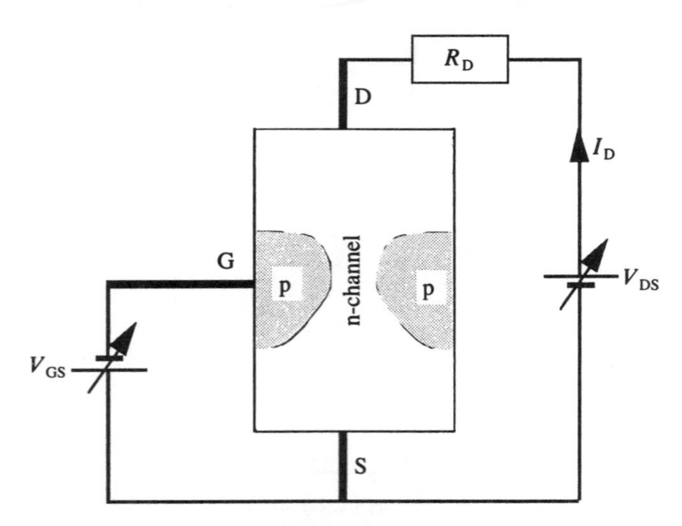

Figure 8.10 *The n-channel JFET biased for operation.*

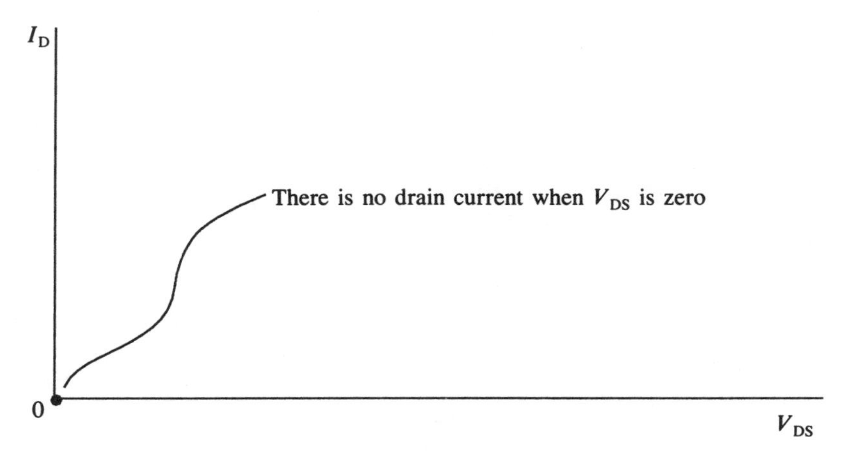

Figure 8.11 *The drain characteristic: when V_{DS} = 0 V, I_D = 0 A.*

V_{GS}, then consider the variation for increasing V_{GS}. Figure 8.10 shows an n-channel JFET biased for operation. Note that V_{DS} and V_{GS} are variable, and that there is a drain resistor R_D to protect the JFET by preventing too high a drain current flowing through the device.

When $V_{GS} = 0$ V

Refer to Figure 8.11. Plot the first point on the graph for $V_{GS} = 0$ V and $V_{DS} = 0$ V: when $V_{DS} = 0$ V a current cannot flow, so $I_D = 0$ A.

Now increase V_{DS}. At first the drain current I_D increases as the bias increases because the depletion region around the gate is not large enough to have a significant effect on I_D. This region is called the ohmic region because the channel resistance $R_{channel}$ is constant, i.e. Ohm's law is obeyed (Figure 8.12).

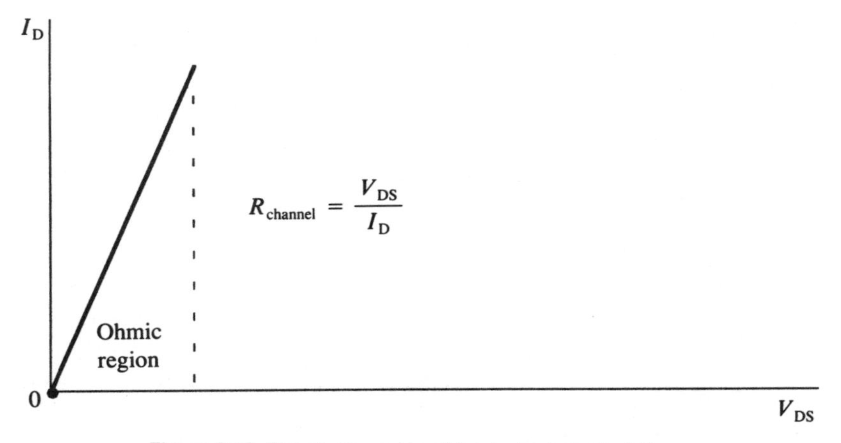

Figure 8.12 *The ohmic region of the drain characteristic.*

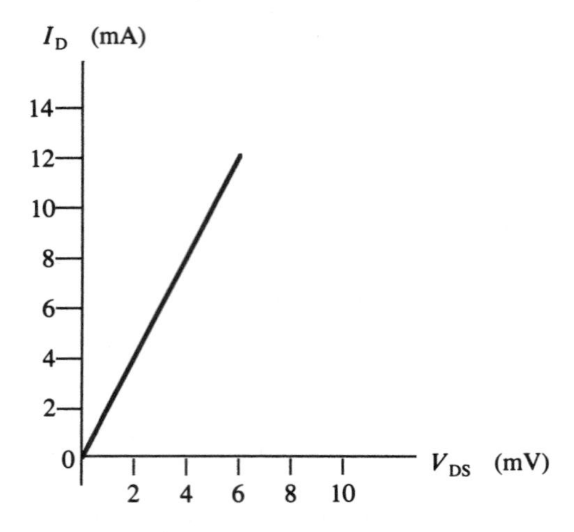

Figure 8.13 *Sketch for Self-assessment Question 8.5.*

SELF-ASSESSMENT QUESTION 8.5

What is the channel resistance of the JFET represented in the Figure 8.13?

As V_{DS} is increased further I_D becomes nearly constant at a value I_{DSS}, the steady-state drain current (Figure 8.14). At this point the reverse-bias voltage across the gate–drain junction V_{GD} causes a depletion region which is sufficient to narrow the channel so that the channel resistance begins to increase significantly. The value of V_{GD} at this point is called the pinch-off voltage V_p—remember it is V_{GD} that causes the narrowing of the channel. As V_{DS} is increased beyond V_p, I_D levels off at I_{DSS}. Figure 8.15 shows the onset on pinch-off in the device.

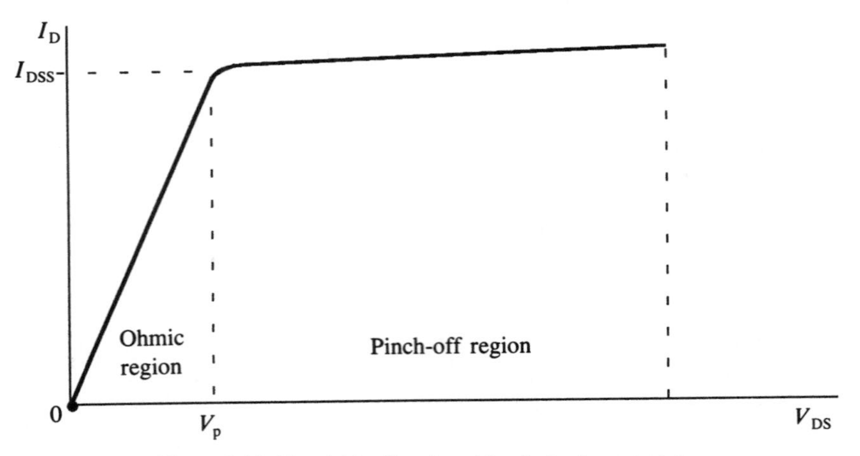

Figure 8.14 *The pinch-off region of the drain characteristic.*

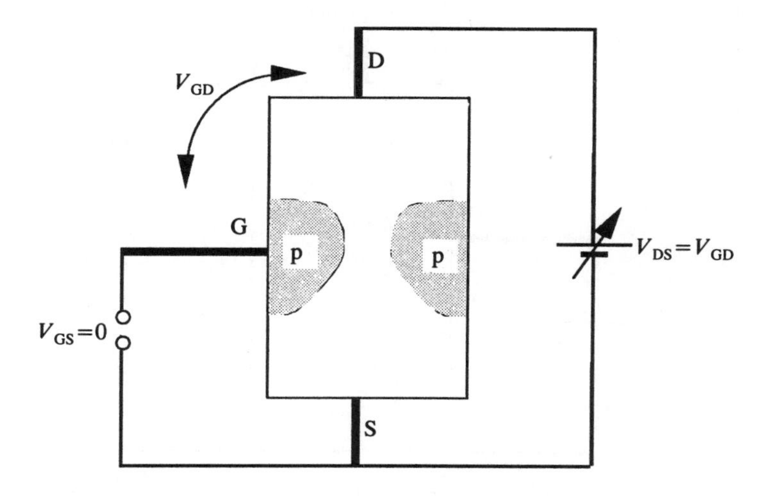

Figure 8.15 *When V_{GD} is sufficiently large to start narrowing the channel, pinch-off occurs ($V_{GD} = V_p$).*

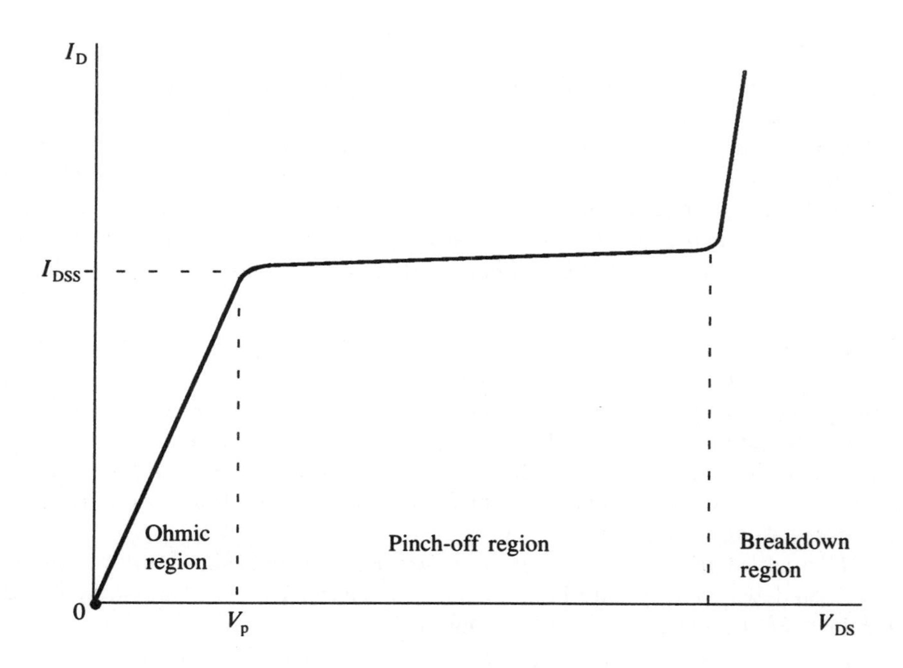

Figure 8.16 *Drain characteristic for an n-channel JFET at $V_{GS} = 0$ V.*

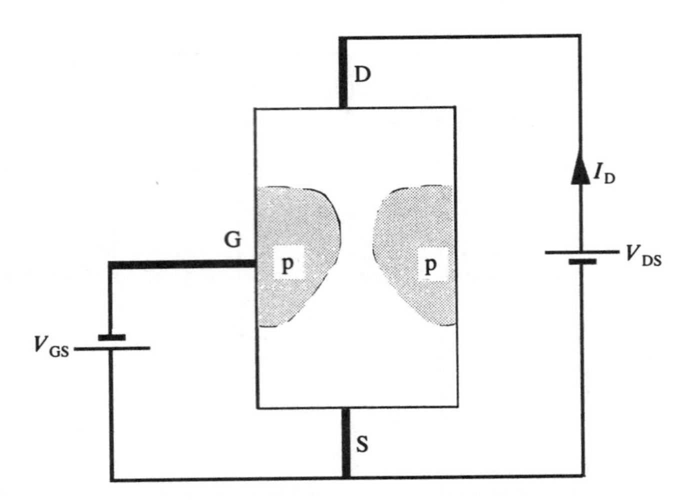

Figure 8.17 *The n-channel JFET with a non-zero V_{GS}: the depletion region around the gate is larger.*

If $V_{GD} = V_p$ and $V_{GS} = 0$, it follows that $V_{DS} = V_p$, i.e. $V_{GD} = V_{DS}$ at pinch-off. In general,

$$V_p = V_{DS(p)} - V_{GS} \tag{8.2}$$

where $V_{DS(p)}$ is the pinch-off value of V_{DS} for a given value of V_{GS}. The value of V_p is constant for a given JFET and represents a fixed parameter, whereas the pinch-off value of V_{DS} (i.e. $V_{DS(p)}$) is a variable which depends on V_{GS}. As V_{DS} is increased further irreversible breakdown occurs and the drain current increases rapidly (Figure 8.16). This can permanently damage the device, so care should be taken to make sure V_{DS} is kept within the operating limit of the device, specified by the manufacturer. In fact, JFETs are operated within the pinch-off region.

Negative V_{GS}

Now consider the case when V_{GS} is no longer zero (Figure 8.17). As V_{GS} is increased from zero (i.e. made more negative), pinch-off occurs at a lower value of V_{DS}. This is because the gate–source bias voltage V_{GS} is more negative so the reverse bias of the gate–source junction is increased. Hence the depletion region around the gate is larger, even when V_{DS} is zero. Channel resistance is therefore increased and the drain current decreases as V_{GS} is made increasingly negative.

The drain characteristic now shows a lower pinch-off voltage, a correspondingly low breakdown voltage, and a larger channel resistance. The drain current I_D will be pinched off at a lower value of I_D (Figure 8.18).

As V_{GS} is changed to increasingly negative values, a family of characteristic curves is obtained (Figure 8.19). The family of curves illustrates the following properties:

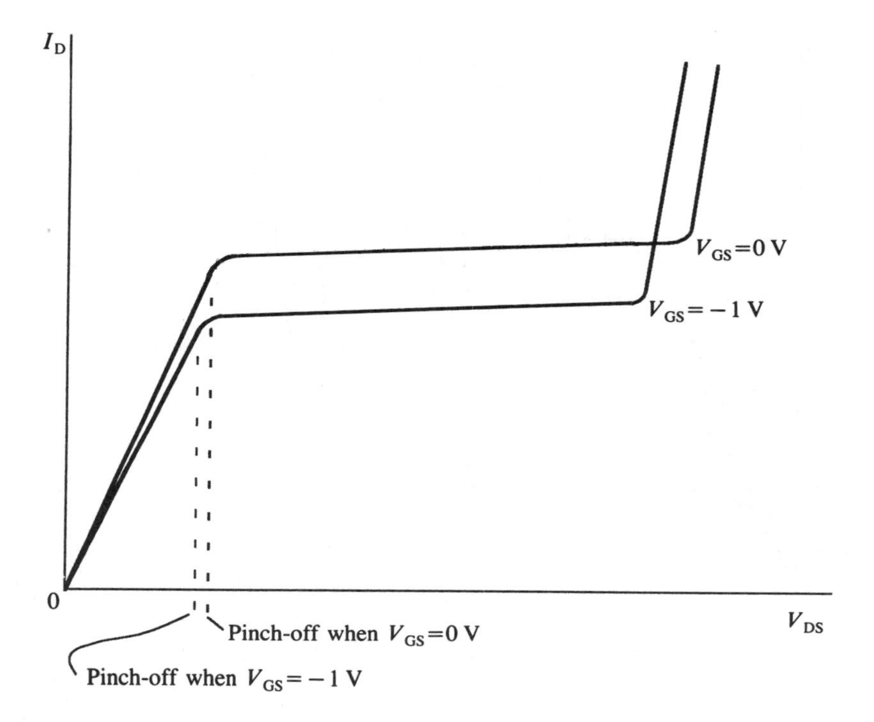

Figure 8.18 *The drain characteristic, showing a lower pinch-off voltage when V_{GS} is increased.*

- increasing gate bias V_{GS} decreases $V_{DS(p)}$;
- increasing gate bias V_{GS} decreases I_D;
- increasing gate bias V_{GS} decreases breakdown voltage;
- maximum drain current I_D occurs for $V_{GS} = 0$ V;
- V_{GS} controls I_D.

What would happen if we were to continue increasing the gate bias V_{GS}? If V_{GS} were made sufficiently negative, would the drain current I_D be reduced to zero? Consider what's actually happening in the JFET to produce the family of curves shown in Figure 8.19. As V_{GS} is made increasingly negative, i.e. as the gate bias is increased, the gate–drain junction is made more and more reverse-biased and the depletion region becomes larger and larger. Hence the channel resistance becomes increasingly larger and the drain current becomes smaller and smaller. Eventually a point should be reached where the depletion region is so large that it meets in the middle of the device (Figure 8.20). At this point no current can get through the channel and we say the FET is cut off. The value of V_{GS} at the cut-off point is called $V_{GS(off)}$.

This cut-off voltage $V_{GS(off)}$ is equal to $-V_p$. We can prove this by considering equation (8.2) as follows:

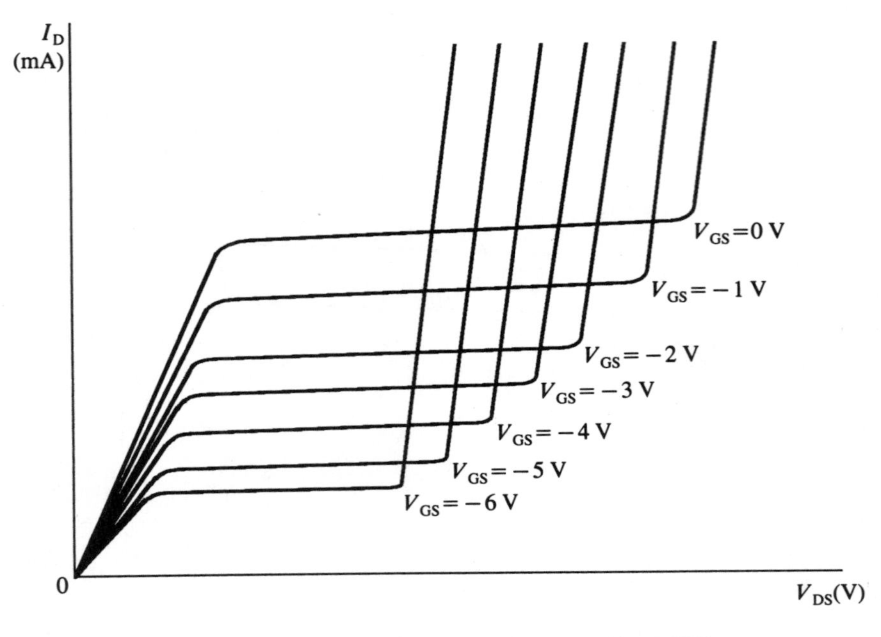

Figure 8.19 *Typical drain characteristics for an n-channel JFET.*

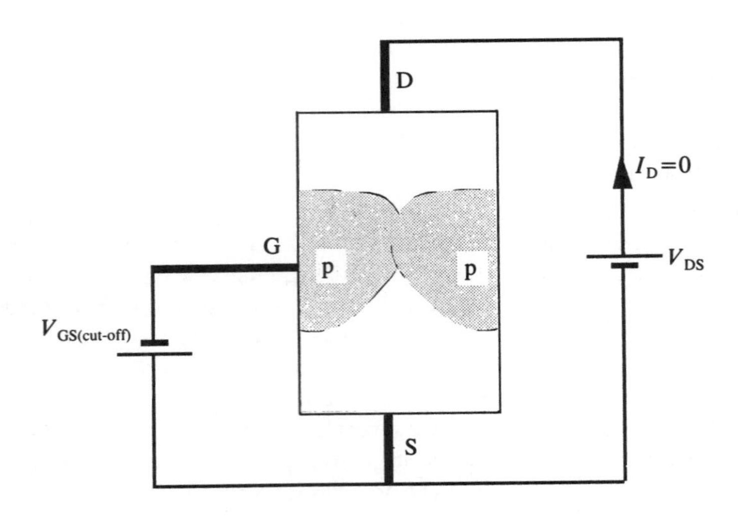

Figure 8.20 *Cut-off in the n-channel JFET.*

If $V_p = V_{DS(p)} - V_{GS}$

When $V_{GS} = 0$ V, $V_p = V_{DS(p)}$

When $V_{GS} = -V_p$, $V_p = V_{DS(p)} + V_p$,

i.e. $V_{DS(p)} = 0$

This means there's no voltage drop between the drain and the source so I_D must be zero. At this point, then,

$$-V_{GS} = V_p \quad \text{or} \quad V_{GS(off)} = -V_p$$

SELF-ASSESSMENT QUESTION 8.6

So what would $V_{GS(off)}$ be for the n-channel JFET illustrated by the set of characteristics shown in Figure 8.21?

Note that I_D is zero only when the magnitude of V_{GS} is equal to or greater than the magnitude of V_p. Generally, $|V_{GS(off)}| = |V_p|$ at the cut-off point. FET data sheets usually give a value for $V_{GS(off)}$ rather than for V_p, but once we know $V_{GS(off)}$ we know V_p, e.g. if $V_{GS(off)} = -4$ V, $V_p = +4$ V.

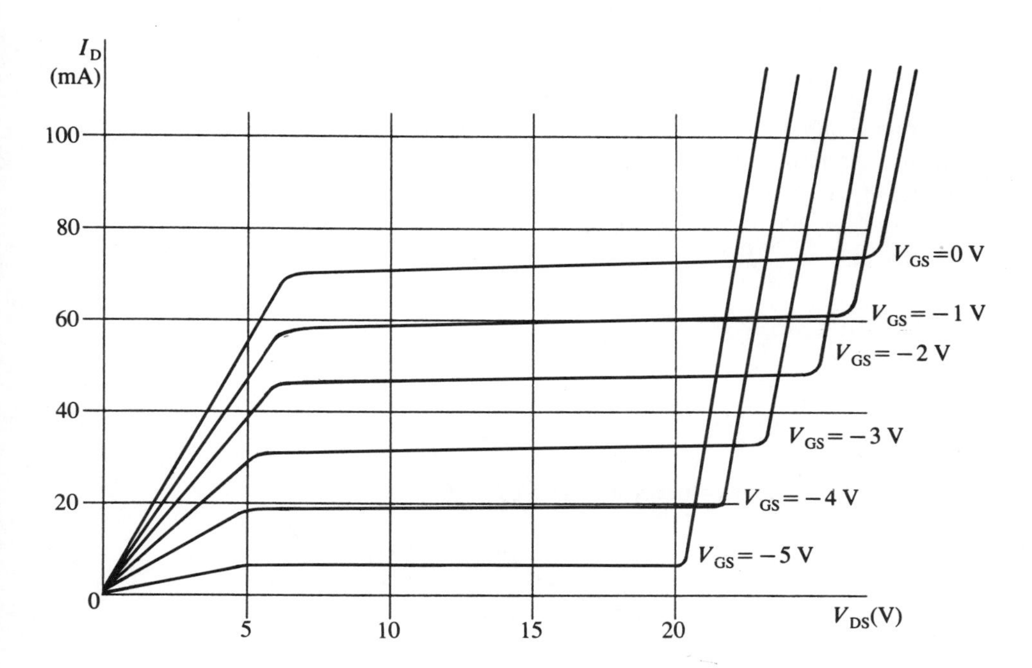

Figure 8.21 *Sketch for Self-assessment Question 8.6 and Self-assessment Question 8.7.*

SELF-ASSESSMENT QUESTION 8.7

Consider the drain characteristics of an n-channel JFET, shown in Figure 8.21. Determine the following:

- channel resistance when $V_{GS} = -1\,V$;
- channel resistance when $V_{GS} = -5\,V$;
- pinch-off voltage V_p;
- steady-state drain current I_{DSS} at $V_{GS} = 0\,V$;
- pinch-off voltage when $V_{GS} = -1\,V$;
- pinch-off voltage when $V_{GS} = -5\,V$;
- breakdown voltage when $V_{GS} = -1\,V$;
- breakdown voltage when $V_{GS} = -5\,V$.

Transfer characteristic and transconductance

The drain characteristic isn't the only way to show the relationships between I_D, V_{DS} and V_{GS}. The transfer characteristic is also used: it shows the relationship between I_D and V_{GS} in the pinch-off region where I_D is almost independent of V_{DS} (Figure 8.22). The transfer characteristic shows very clearly the operating limits of the JFET: $I_D = 0$ when $V_{GS} = V_{GS(off)}$ and $I_D = I_{DSS}$ when $V_{GS} = 0$. The transconductance g_m is an important FET parameter because it determines voltage gain in FET amplifiers. Transconductance is equal to the slope of the transfer characteristic:

$$g_m = \Delta I_D / \Delta V_{GS}$$

From Figure 8.22 you can see that g_m is larger near $V_{GS} = 0$ than near $V_{GS(off)}$. Manufacturers usually specify g_m measured at $V_{GS} = 0\,V$ (called g_{m0}). There are certain other parameters that manufacturers specify for JFETs. These include V_p, $V_{GS(off)}$, I_{DSS}, and $V_{breakdown}$ for $V_{GS} = 0\,V$.

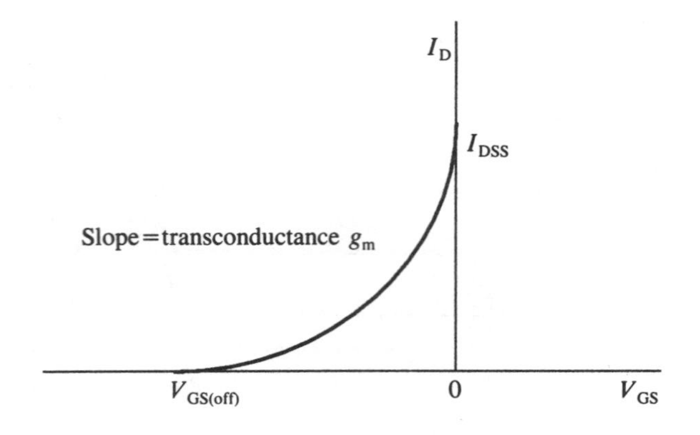

Figure 8.22 *Transfer characteristic of an n-channel JFET.*

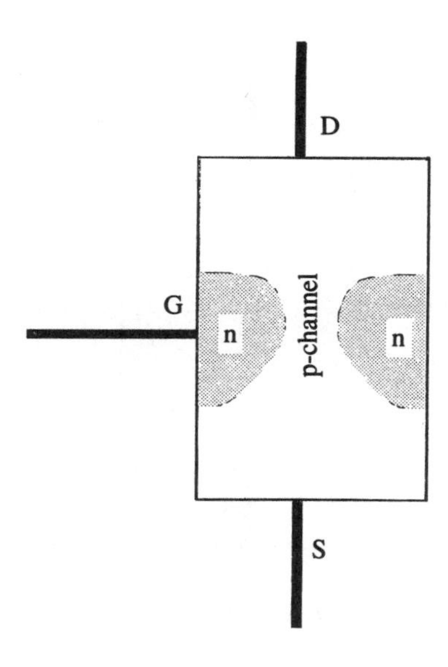

Figure 8.23 *Sketch for Self-assessment Question 8.7.*

8.2.1 ***Operation of the p-channel JFET***

n-type FETs are used more frequently than p-type FETs because of the higher mobility of electrons. Before the operation of the p-channel JFET is described, try and work out some of it yourself.

SELF-ASSESSMENT QUESTION 8.8

Consider the p-channel JFET shown in Figure 8.23. Can you work out how the two bias voltages V_{GS} and V_{DS} should be applied? Remember that both the gate–drain junction and the gate–source junction must be reverse biased.

The principles of operation of the p-channel JFET are the same as those for the n-channel JFET. This means that V_{DS} is used to provide a current flowing through the device, such that the source is a source of the majority carrier. For the p-channel JFET this means the source must 'send' holes upwards through the device, meaning that I_D must flow through the device from source to drain, unlike the n-channel device where I_D flows from drain to source. V_{GS} is used to control the size of I_D by controlling the size of the depletion region around the gate, so the gate–source junction must be reverse-biased so that the depletion region can be made very large to enable the eventual cut-off of the current (Figure 8.24).

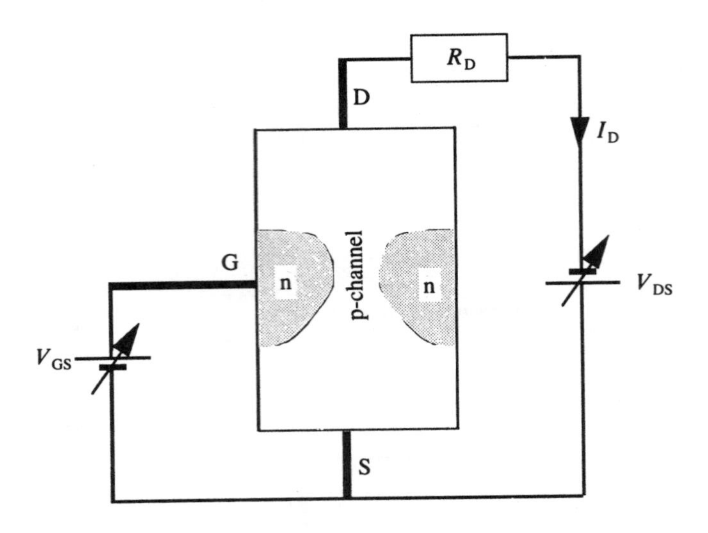

Figure 8.24 *A p-channel JFET biased for operation.*

p-channel JFET drain characteristic

Increasing the gate bias voltage (i.e. making V_{GS} increasingly positive) makes the depletion region around the gate larger, leading to a reduction in I_D. Eventually the depletion region can be made so large that cut-off is obtained. Following the same reasoning as that applied to an n-channel JFET, the p-channel drain characteristic of Figure 8.25 is obtained. Note that values of V_{DS} are negative. p-channel JFET drain characteristics are usually shown plotted the other way around to resemble the previous n-channel drain characteristics (Figure 8.19).

8.2.2 JFET structure

So far we've talked about the JFET as though it's a rod of semiconductor. It is possible to manufacture FET's in this form, but these days integrated-circuit transistors are more common ('chips'). Planar technology is used to manufacture integrated circuits which may contain thousands of transistors and other electronic devices. The different regions of the semiconductor device are deposited onto a substrate (called a wafer) in the form of layers in such a way that thousands of each device can be grown simultaneously. Further processing is carried out to deposit the metal electrodes and insulating layers such as silicon dioxide, then the wafer is sawn into individual chips. Figure 8.26 is a sketch of a planar n-channel JFET. It has a sandwich structure with the electrodes all on the upper surface.

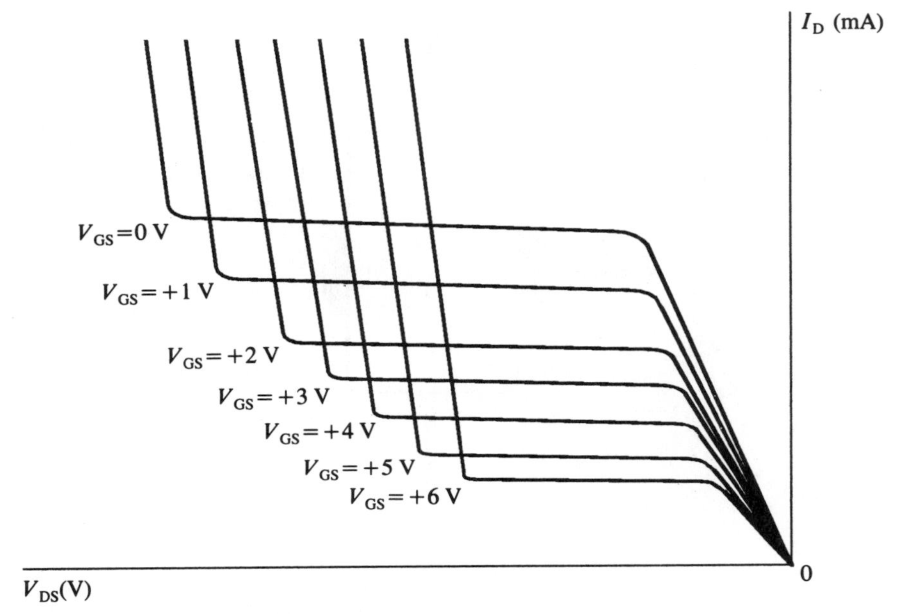

Figure 8.25 *Drain characteristic of a p-channel JFET.*

8.3 The MOSFET

MOSFETs are very important devices: they're used extensively in modern integrated circuits for microprocessors and computer memories. They differ structurally from JFETs in that a layer of insulator is inserted between the gate electrode and the gate region in the semiconductor (Figure 8.27). The most commonly

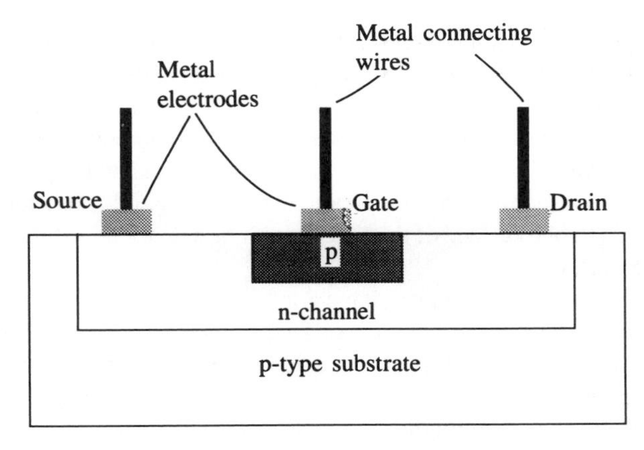

Figure 8.26 *A planar n-channel JFET.*

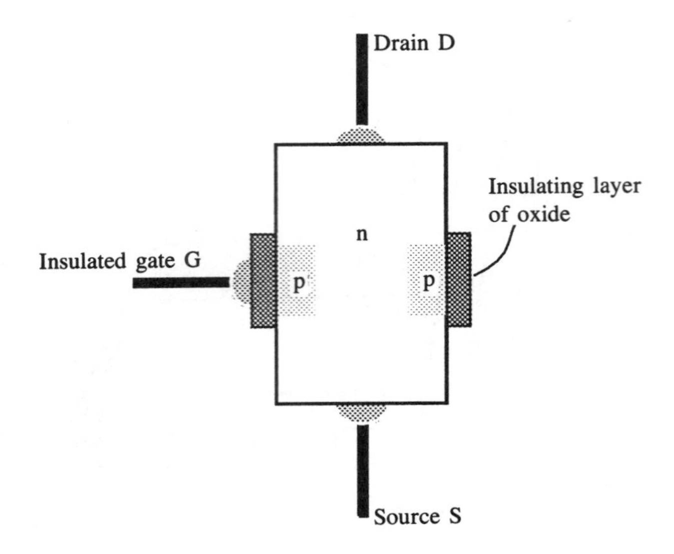

Figure 8.27 *Schematic representation of a MOSFET.*

used insulator is an oxide, silicon dioxide SiO_2, and the most common semiconductor used is silicon. Hence the name MOSFET: *m*etal–*o*xide–*s*emiconductor FET, also known as the IGFET (*i*nsulated-*g*ate FET), the MISFET (*m*etal–*i*nsulator–*s*emiconductor FET) and the MOST (*m*etal–*o*xide–*s*emiconductor *t*ransistor). Commonly used symbols for the MOSFET are shown in Figure 8.28. There are two types of MOSFET: *enhancement* and *depletion–enhancement* (sometimes just called *depletion*).

8.3.1 **Enhancement MOSFET operation**

The operating principles of the MOSFET are similar to those of the JFET, but the situation is made more complicated by the layer of oxide between the metal electrode and the semiconductor. The metal electrode referred to is the one at the gate. The sketch of Figure 8.29 represents a planar n-channel enhancement MOSFET. Note that the enhancement MOSFET does not have a channel fabricated into the device.

It's important to remember that the oxide layer is insulating. This means that any charge transferred to one surface of the oxide won't be able to move through the oxide to the opposite surface.

Now suppose that a positive gate voltage is applied to the MOSFET of Figure 8.30. This positive voltage charges up the top surface of the oxide layer so that a layer of positive charge is lying at the top of the oxide. This layer of positive charge then attracts electrons by induction in the semiconductor near the oxide such that the electrons move to the semiconductor–oxide interface. These electrons are the minority carriers in the semiconductor.

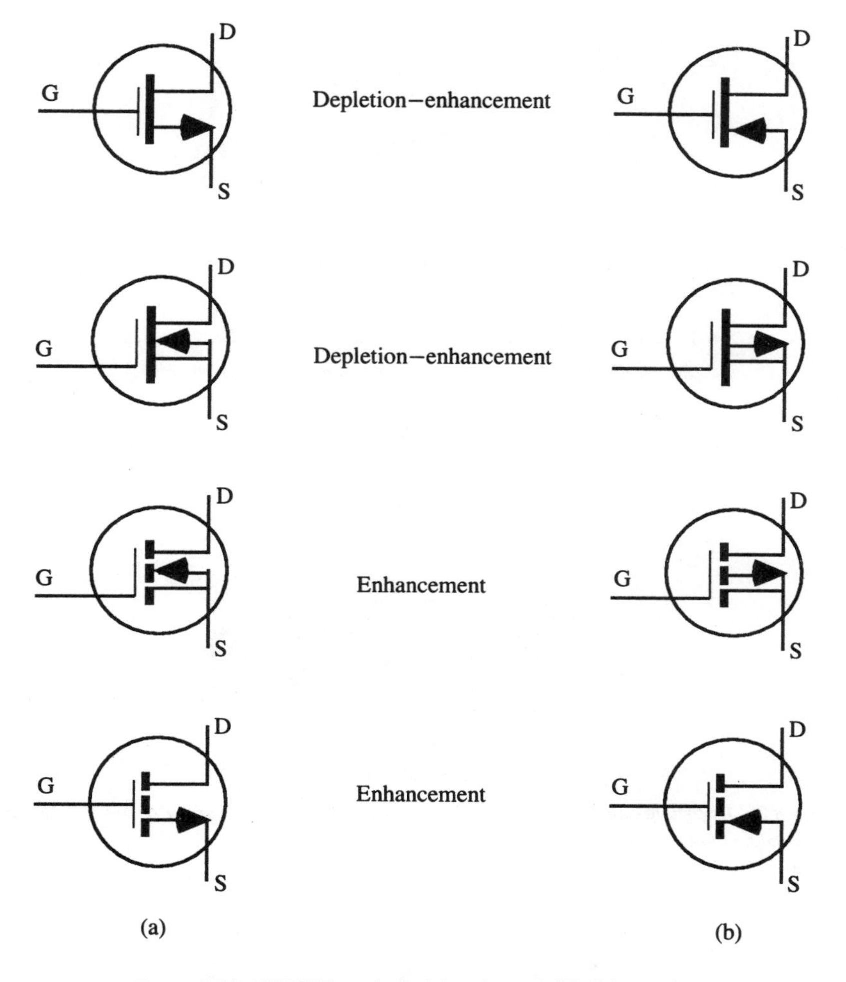

Figure 8.28 *MOSFET symbols: (a) n-channel, (b) p-channel.*

This electron layer is important because it connects the n-type regions of the source and the drain, making a channel through which electrons could flow between them (Figure 8.31). Hence, even though the semiconductor is p-type the channel is n-type. A certain threshold gate voltage $V_{GS(th)}$ is required before the channel is induced, typically less than 4 V.

Now apply a positive voltage V_{DS} to the drain such that the source is relatively negative (Figure 8.32). As in the case of the JFET, this can be achieved by grounding the source. When a small drain voltage is applied, a small drain current I_D flows from the drain to the source (electrons flow from the source to the drain). Hence a plot of

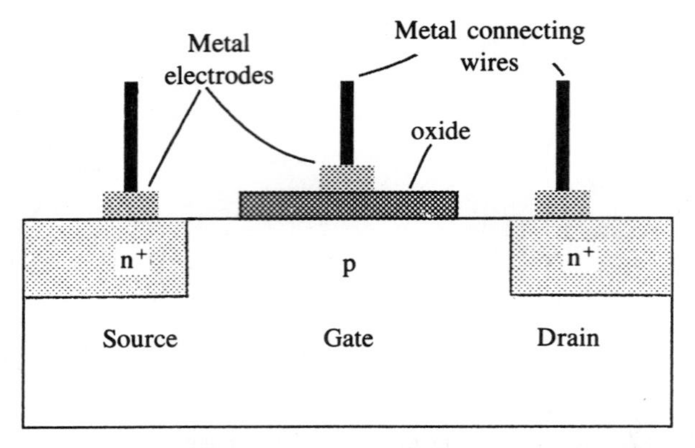

Figure 8.29 *Planar structure of an n-channel enhancement MOSFET.*

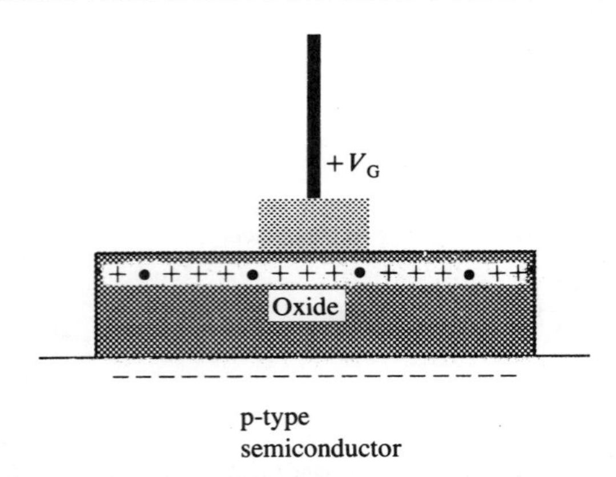

Figure 8.30 *A positive gate voltage applied to the gate in the MOSFET, showing the induced-charge layers each side of the oxide.*

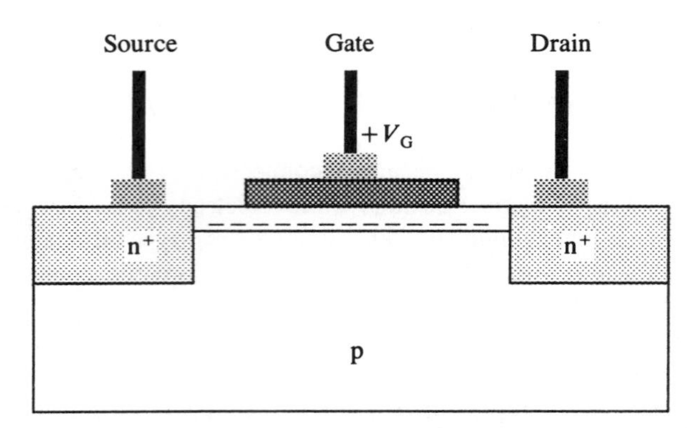

Figure 8.31 *The formation of an n-channel in the MOSFET by applying a positive gate voltage.*

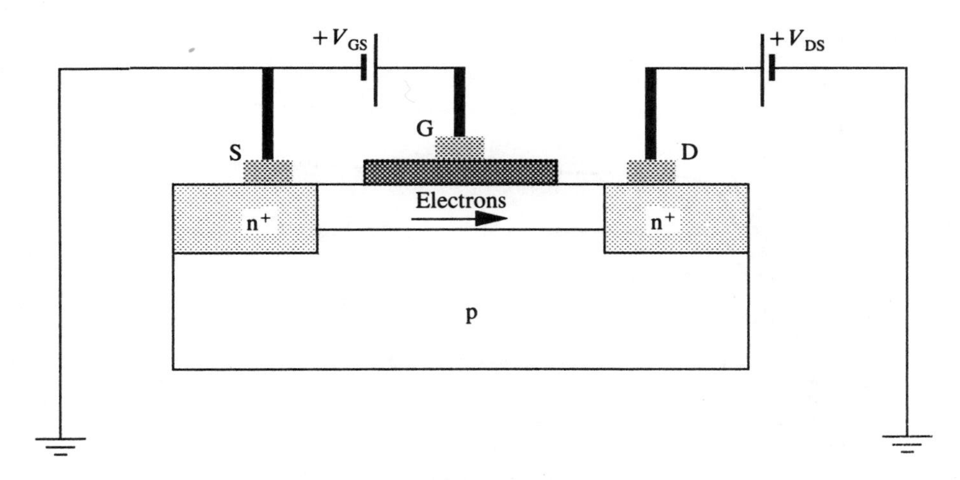

Figure 8.32 *MOSFET biased for conduction of electrons through the n-channel.*

drain current I_D versus drain voltage V_{DS} produces a straight line as in Figure 8.33, the slope of which gives the channel resistance, as in the case of the JFET.

As V_{GS} is increased beyond $V_{GS(th)}$ the density of electrons in the channel increases, resulting in an increased drain current. This is opposite to the JFET case where increasing V_{GS} results in decreased drain current.

As V_{GS} is increased, the amount of positive charge on the oxide is increased leading to an increase in the number of electrons moving into the induced n-channel and the channel eventually becomes saturated. At this point I_D levels off and becomes almost constant, as in the JFET case.

It is the *enhancement* of the channel by increasing V_{GS} that gives this particular MOSFET its name. A family of drain characteristics (Figure 8.34) is obtained by varying V_{DS} and measuring I_D for different values of V_{GS}. The transfer characteristic can then be plotted (Figure 8.35).

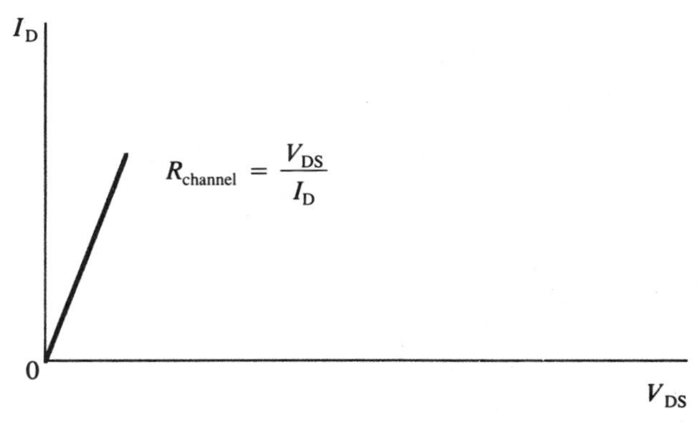

Figure 8.33 *The ohmic drain characteristic for low values of V_{DS} for an enhancement MOSFET.*

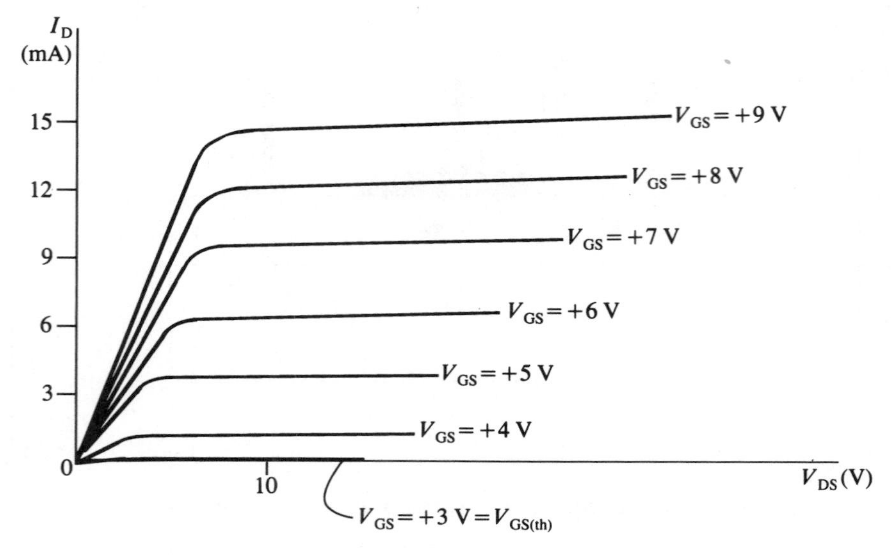

Figure 8.34 *Drain characteristics for a typical n-channel enhancement MOSFET.*

8.3.2 **Depletion–enhancement MOSFET operation**

In the depletion MOSFET the channel is built into the device; it is not induced. Figure 8.36 shows a sketch of a simple depletion MOSFET with an n-channel between the n-type source and drain. This type of MOSFET can be used either in enhancement mode or depletion mode.

If a positive gate voltage is applied channel conductivity is enhanced by electrons being attracted into the channel from the p-type substrate, and the situation is analogous to that for the enhancement MOSFET described previously.

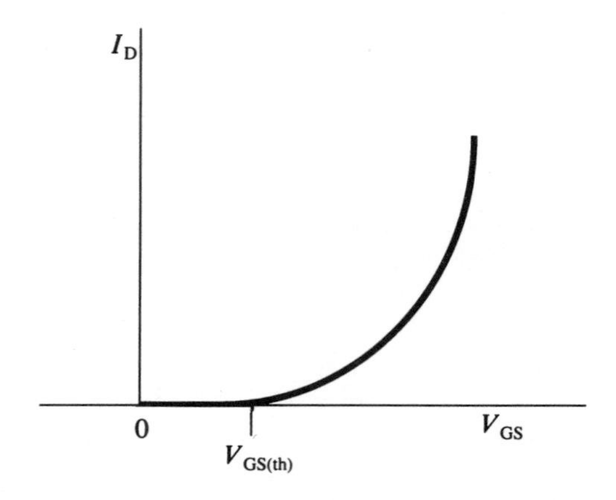

Figure 8.35 *Transfer characteristic for an n-channel enhancement MOSFET.*

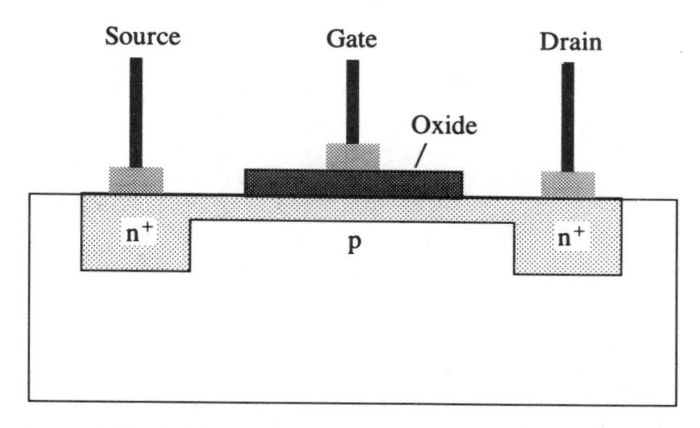

Figure 8.36 *Planar structure of an n-channel depletion MOSFET.*

If the applied gate voltage is made negative relative to the source, holes (majority carriers) in the p-type semiconductor will move into the n-channel (Figure 8.37). This process leads to recombination in the channel and the resulting drain current I_D is reduced. The greater the negative gate bias, the greater the rate of recombination leading to lower values of I_D. This is called *depletion* and gives similar characteristics to the n-channel JFET. Figure 8.38 shows the transfer characteristic for an n-channel depletion–enhancement MOSFET operated in the depletion mode.

8.4 Summary

Field-effect transistors are three-terminal devices which produce a current between the drain and the source which is controlled by a voltage at the gate. The current flows

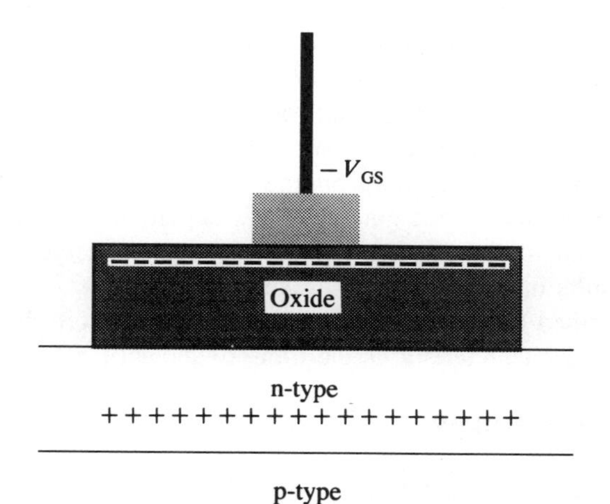

Figure 8.37 *A negative gate voltage applied to the gate in the depletion MOSFET.*

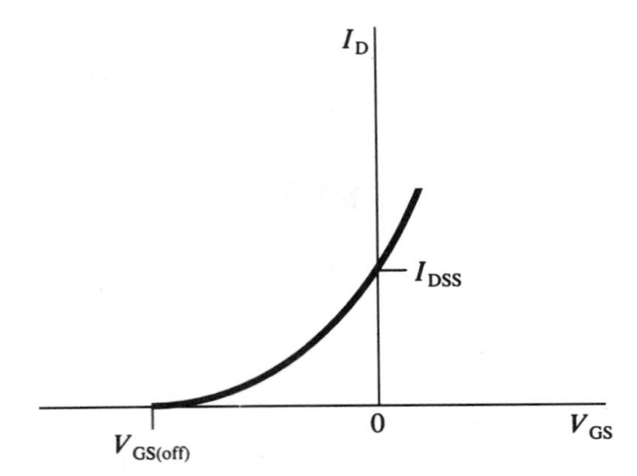

Figure 8.38 *Transfer characteristic for the n-channel depletion MOSFET.*

through a channel which is either n-type or p-type in character. Plotting the size of the current in the channel (called the drain current I_D) against the voltage applied between the drain and source V_{DS} produces a family of characteristic curves for varying values of gate–source voltage V_{GS}. FETs are operated in the pinch-off (saturation) region of the drain characteristic.

The operation of the JFET depends on the size of the depletion region around the reverse biased gate: when V_{DS} is increased to V_p (or $-V_{GS(off)}$) the current saturates and the device is said to be pinched off, at which point the current reaches its steady-state value I_{DSS}. If the reverse bias V_{GS} is increased to $V_{GS(off)}$ the channel is cut off and the drain current decreases to zero.

The operation of the enhancement MOSFET depends on the existence of an induced channel below the insulated gate. The channel is induced between source and drain in the substrate material. The channel is induced when a sufficient positive voltage (for an n-channel device) is applied to the gate; the threshold voltage at which this occurs is $V_{GS(th)}$. Electrons then form a current through the channel. This drain current I_D increases with increasing gate voltage V_{GS}.

The depletion–enhancement MOSFET has a channel between the source and the drain. Positively biasing the gate (for an n-channel device) causes enhancement: electrons are drawn into the channel from the p-type substrate and the drain current increases. Negatively biasing the gate produces depletion in the channel when the majority carriers from the substrate (holes, for an n-channel device) recombine in the channel. This results in a decreased drain current.

FETs are described by drain characteristics and by transfer characteristics. The slope of the transfer characteristic is the transconductance g_m.

8.5 **Tutorial questions**

8.1 A particular p-channel JFET has a $V_{GS(off)}$ value of 4 V. What is I_D when $V_{GS} = 6$ V?

8.2 The drain–source voltage at the pinch-off point of a particular JFET is 7 V. If the gate–source voltage is zero, what is V_p?

8.3 The V_{GS} of a certain n-channel JFET is increased negatively. Does the drain current increase or decrease, and why?

8.4 What value must V_{GS} have to produce cut-off in a p-channel JFET with a V_p of 3 V?

8.5 Draw a sketch to show how you would bias an n-channel JFET for operation. Explain how the drain characteristic of an n-channel JFET is formed. Use sketches to illustrate your answer.

8.6 Using your knowledge of JFETs and the n-channel MOSFET, describe in your own words how you would expect a p-channel depletion–enhancement MOSFET to operate in both enhancement and depletion modes. Sketch the transfer characteristics.

8.6 Suggested further reading

Boylestad, R. and Nashelsky, L., *Electronic Devices and Circuit Theory*, 5th edn, Prentice Hall, 1992, pp. 207–239.

Floyd, T.L., *Electronic Devices*, 3rd edn, Merrill, 1992, pp. 292–304, 316–323.

CHAPTER 9

Bipolar transistors and their operation

Aims and objectives

This chapter discusses the bipolar junction transistor, or BJT, which was invented in 1948 by three men who shared the 1956 Nobel Prize for Physics for their invention : John Bardeen, Walter Brattain and William Shockley.

In this chapter the operation of the pnp and npn BJT will be described in terms of the transistor's two p–n junctions and the current flow across them. Bias conditions will be described and a circuit given for the plotting of common-emitter collector characteristics. Important BJT parameters such as common-base current gain α and common-emitter current gain β will be defined and equations derived relating these to each other and to the three terminal currents in the device.

9.1 Structure of the BJT

The operation of a BJT involves both holes and electrons, hence the name *bipolar*.

As with the FET, the operation of the bipolar junction transistor (BJT) depends on the existence of p–n junctions within the transistor. A BJT consists of three semiconductor regions which are separated from each other by p–n junctions (Figure 9.1). Each separate region has a metal electrode attached to connect the device to a circuit. In the BJT the three regions are always called:

- collector
- emitter
- base

The emitter and the collector are the same type (i.e. both n-type or p-type), and the base is the opposite type. The base region is narrow and lightly doped compared to the collector and emitter regions. There are two main types of BJT: the pnp and the npn. Sketches of the structures of these with their circuit symbols are shown in Figure 9.2. Note the directions of the arrowheads in the circuit symbols.

9.2 Operation of the BJT

First, consider the properties of a p-n junction under forward and under reverse bias (you may wish to refer back to Chapter 5 to refresh your memory). Diffusion is by majority carriers, drift is by minority carriers. Remember the forward current and the reverse current both consist of a diffusion and a drift component: in forward bias the

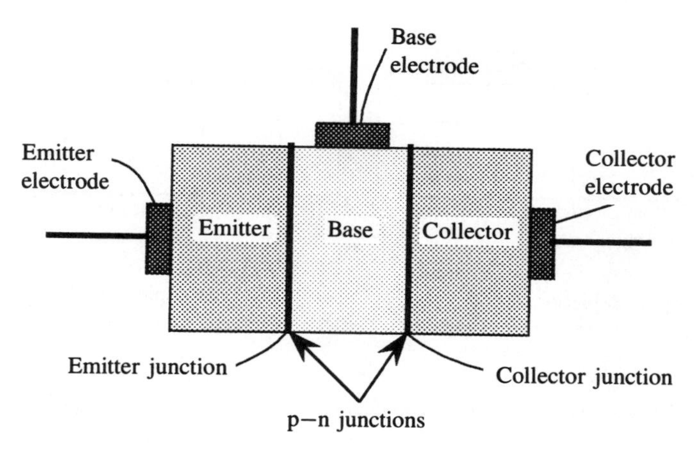

Figure 9.1 *Basic structure of a bipolar junction transistor.*

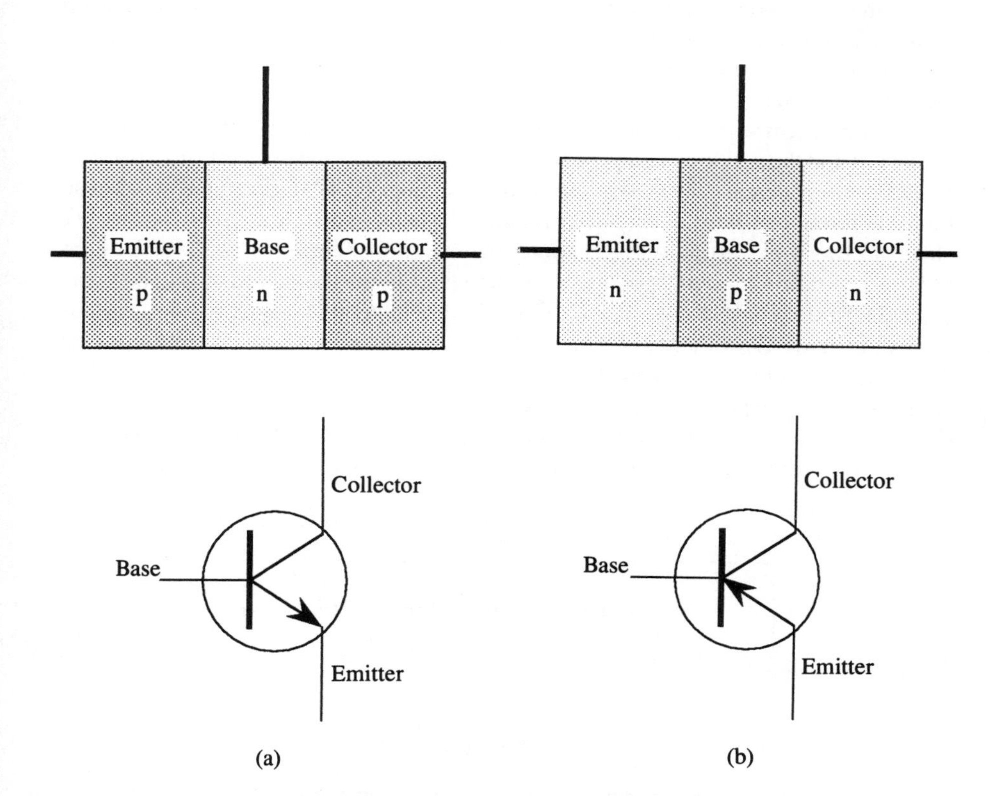

Figure 9.2 *Schematic diagrams of (a) the npn bipolar junction transistor and (b) the pnp bipolar junction transistor, with their circuit symbols.*

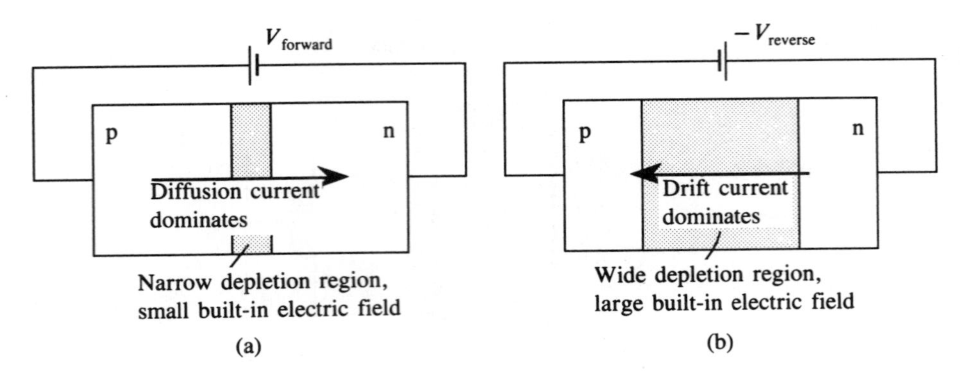

Figure 9.3 *The p–n junction under (a) forward and (b) reverse bias.*

diffusion component is dominant and the depletion region is narrow; in reverse bias the depletion region is wide and the diffusion component is negligible so that drift, although small, is dominant (Figure 9.3). The reverse saturation current I_o is made up almost entirely of drifting carriers. The majority-carrier concentration on each side of the p–n junction varies with the applied bias because of variations in the diffusion of carriers across the junction.

Now consider the reverse-bias p–n junction. I'm going to re-draw it with the n-side on the left and the p-side on the right because this will make the description of the BJT easier later on (Figure 9.4). Any electron–hole pairs (EHPs) generated will cause minority holes to drift from n to p and minority electrons to drift from p to n (remember electrons travel in the opposite direction to conventional current). So the size of the reverse saturation current I_o depends on the generation of electron–hole pairs: the more EHPs there are, the greater I_o. If an EHP is generated on the n-side near the depletion region, the minority hole created has a good chance of being swept across the depletion region by the built-in electric field providing the hole is within a diffusion length L_p of the depletion region. Hence a minority-hole current flows from n to p (Figure 9.5). The majority electron which is the other partner in the EHP forms part of the diffusion current in the device. If we were able to increase the rate at which EHPs

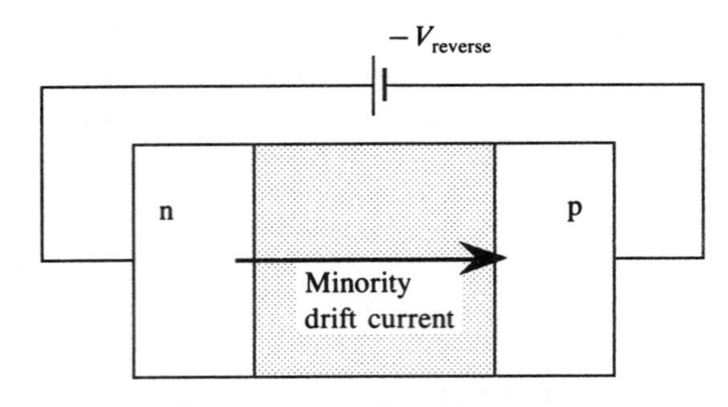

Figure 9.4 *The reverse-bias p–n junction redrawn.*

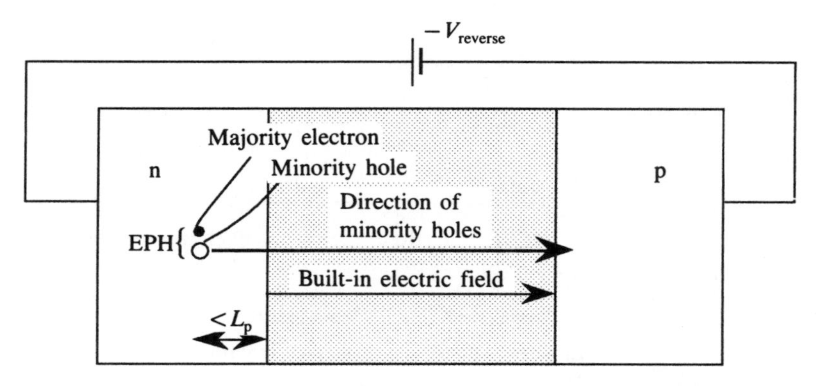

Figure 9.5 *Generation and transport of a minority hole within a diffusion length L_p of the deplection region.*

are generated on the n-side, then the size of the minority-hole current (and therefore I_o) would increase.

SELF-ASSESSMENT QUESTION 9.1

What are meant by diffusion length and lifetime? (If you can't remember, refer to Chapter 4).

This process is easily carried out in a photodiode (see Section 7.4) where optical excitation is used to generate the excess EHPs. If we could 'inject' many EHPs by electrical means instead, then all the minority holes so created would be swept across the depletion region into the p-side, providing the EHPs were generated within L_p of the depletion region. We could thereby produce a large reverse saturation current. The hole current from n to p would then depend on the rate of hole injection into the n-side within L_p of the depletion region.

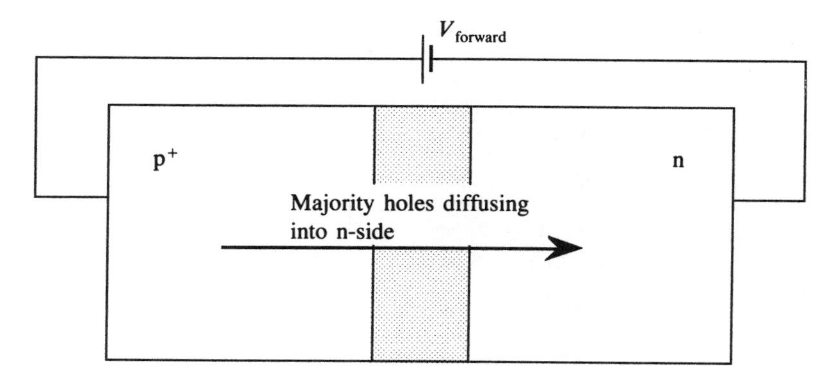

Figure 9.6 *Forward-biased p^+n junction producing many injected holes in the n-side.*

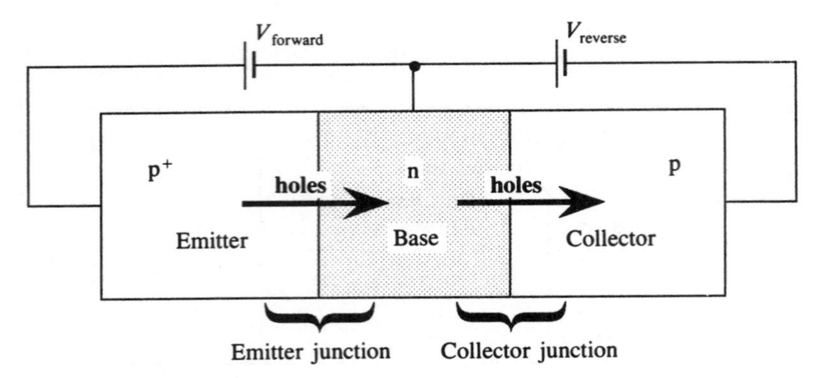

Figure 9.7 *Operating principle of a pnp bipolar junction transistor.*

Now consider a forward-biased p^+n junction. Note the + sign after the p: this means the p-side is heavily doped compared to the n-side. A forward-biased p^+n junction can be used to inject lots of holes from the p-side into the n-side by diffusion (Figure 9.6).

If we make the n-side of this forward-biased p^+n junction the n-side of the previous reverse-biased p–n junction, we obtain a single device containing two p–n junctions (Figure 9.7). This is a pnp bipolar junction transistor. The heavily doped p^+ region is called the emitter because it injects many holes into the adjoining region by diffusion across the emitter junction. The middle n region is called the base. The right-hand p region is called the collector because it collects holes from the emitter. The holes in the base will be swept into the collector by the built-in electric field across the reverse-biased collector junction. Figure 9.7 shows a pnp BJT in the common-base configuration. This is so-called because the base is common to both the emitter circuit and the collector circuit.

There are three design requirements which should be explained before we go any further.

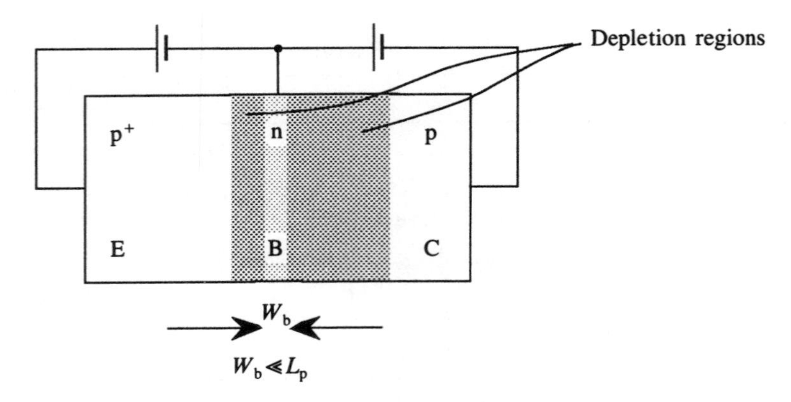

Figure 9.8 *Base width W_b must be narrow compared to diffusion length L_p.*

1. The width of the base W_b must be significantly less than the hole diffusion length L_p (Figure 9.8). This is so that holes injected into the base from the emitter will have a good chance of reaching the collector without recombining.
2. The holes in the base must have a long lifetime τ_p to prevent them recombining before they have a chance of reaching the collector.
3. Recombination between diffusing holes and electrons across the emitter junction must be minimized. This can be achieved by heavy doping in the emitter so that the majority-carrier concentration in the emitter is much larger than the majority-carrier concentration in the base. In the pnp BJT, therefore, the current crossing the emitter junction will consist mostly of holes going from the emitter to the base, rather than electrons going from the base to the emitter. So dope the base lightly and dope the emitter heavily:

$$N_{\mathrm{aEmitter}} \gg N_{\mathrm{dBase}}$$

9.2.1 **Current flow in the pnp BJT**

First, consider the two p–n junctions in the pnp BJT.

SELF-ASSESSMENT QUESTION 9.2
How many current components will there be across each junction in the pnp BJT? What are they?

We know that the hole diffusion current across the emitter junction of the pnp BJT will be large, because the emitter is heavily doped and because the emitter junction is

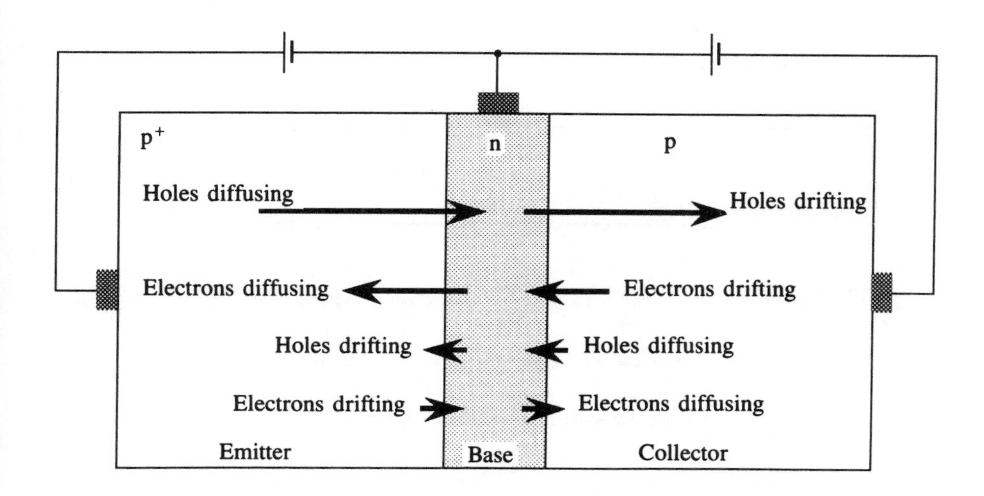

Figure 9.9 *Carrier flow across the emitter and collector junctions in the pnp BJT.*

forward biased. We also know that there will be an electron diffusion current across the same junction, but this will be smaller than the hole component. There will also be two drift components across this junction: the hole drift current and the electron drift current; these will both be insignificant because the junction is forward biased. Figure 9.9 illustrates the carrier flow across this junction, and Figure 9.10 shows the current flow.

The situation is different at the collector junction because it is reverse-biased. Here both diffusion components are negligible whereas the drift components are significant. Hole drift is particularly important — remember the pnp BJT is designed so that lots of holes are injected from the emitter into the base. If the base region is narrow these holes will be swept across the collector junction by the built-in electric field there, i.e. there will be a large hole drift current across the collector junction. There will also be an electron drift current but it won't be as large as the hole component because there are fewer minority electrons on the p-side (collector) than there are minority holes on the n-side (the base). See Figures 9.9 and 9.10.

SELF-ASSESSMENT QUESTION 9.3

Why are there fewer minority electrons in the collector than minority holes in the base?

From the discussion so far you should be able to work out which are the significant currents across the emitter and collector junctions. Remember that, according to the model developed in this book, there are always four current components across each junction. Some of these are significantly larger than others, meaning that some of the components can be neglected.

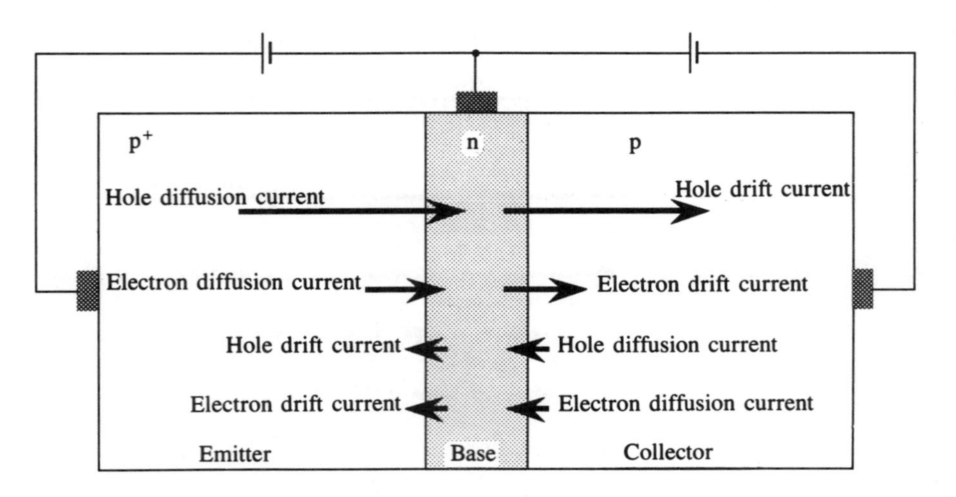

Figure 9.10 *Current flow across the emitter and collector junctions in the pnp BJT.*

SELF-ASSESSMENT QUESTION 9.4

From the previous discussion which emitter and collector current components would you say can be neglected in the pnp BJT?

We must now consider the base in the pnp BJT. Firstly, some holes entering the base from the emitter will recombine with electrons in the base. The BJT is designed to minimize this recombination, but some will occur. Secondly, electrons will diffuse from the base into the emitter (remember the emitter junction is forward-biased). Thirdly, thermally generated EHPs in the collector will mean electrons drift from the collector into the base (remember the collector junction is reverse-biased). These three carrier flows are represented in Figure 9.11.

These current components in the base are small. The base is narrow so there are relatively few dopant atoms there to donate electrons, compared with the number of dopant atoms in the emitter and the collector. Remember the emitter and collector regions are much wider than the base. As there are relatively few electrons in the base, therefore, they will have to be supplied from the external circuit via the wire attached to the base electrode. Even though the currents in the base are small, the existence of the base current is very important because it provides the basis for current gain (amplification) in the BJT. The base current in the pnp BJT leaves the transistor and enters the external circuit — remember conventional current moves in the opposite direction to the electrons.

The three terminal currents in the pnp BJT are shown in Figure 9.12. The emitter current is I_E, the collector current is I_C, and the base current is I_B. Note the direction of current flow: the emitter current enters at the emitter, the collector current leaves the collector, and the base current leaves the base. Note that most of the current in the pnp BJT consists of the holes injected by the emitter: most of these pass through the base

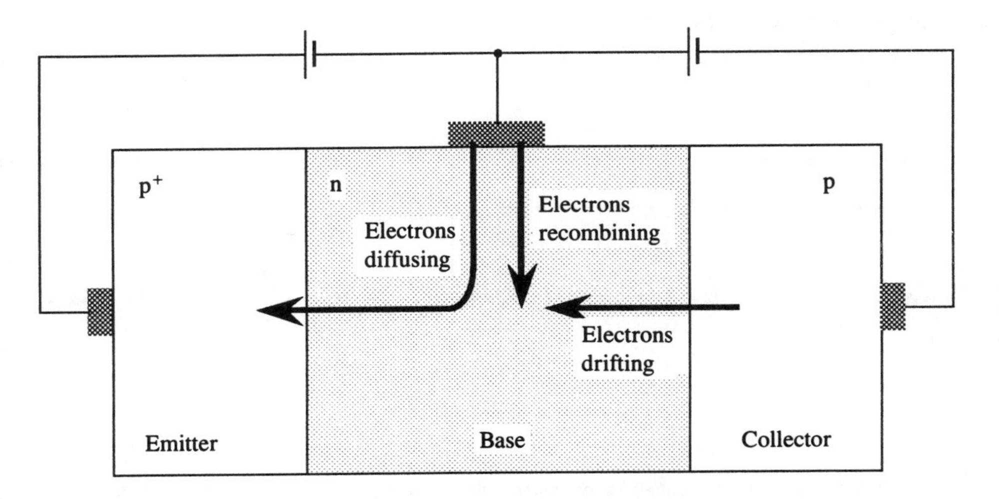

Figure 9.11 *Carrier flows in the base of the pnp BJT.*

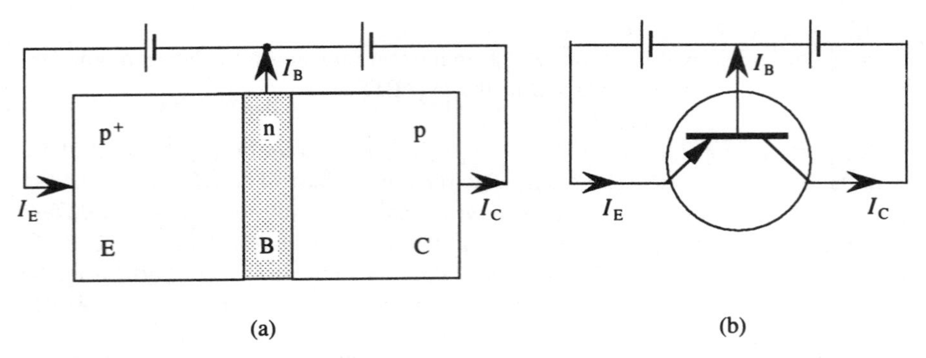

Figure 9.12 *Terminal currents in (a) the pnp bipolar junction transistor and (b) represented on the pnp symbol.*

into the collector and so form most of the collector current. Observe the direction of the arrowhead on the emitter arm of the pnp symbol in Figure 9.12—it's in the direction of the emitter current.

9.2.2 Current flow in the npn BJT

The operating principles of the npn BJT are identical to those of the pnp BJT except that the base is now p-type, and the emitter and collectors are n-type. This effectively reverses the biasing required across the emitter and collector junctions, and reverses the current directions (Figure 9.13). Most of the current in the npn BJT consists of the electrons injected by the emitter: most of these pass through the base into the collector and so form most of the collector current. Note the direction of the arrowhead on the emitter arm of the npn symbol in Figure 9.13. As in the pnp case, it's in the direction of the emitter current.

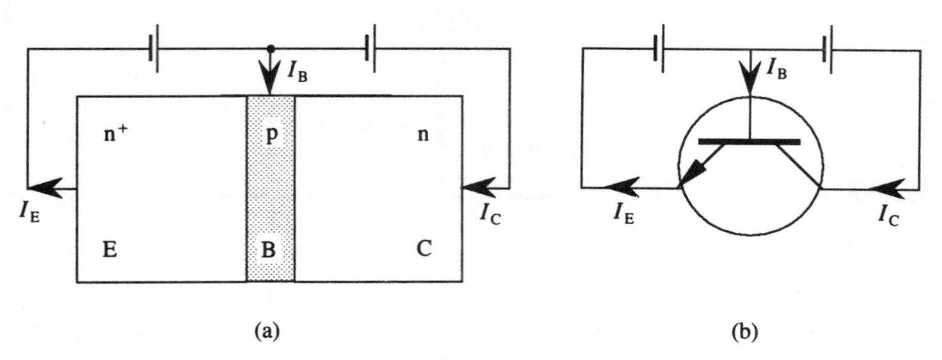

Figure 9.13 *Terminal currents in (a) the npn bipolar junction transistor and (b) represented on the npn symbol.*

9.3 **BJT parameters**

We're now in a position to write some equations describing BJT operation. We can write equations for the total terminal currents I_E, I_C and I_B:

$$I_E = I_{Ep} + I_{En} \tag{9.1}$$

where I_{Ep} is the hole component of the emitter current and I_{En} is the electron component of the emitter current;

$$I_C = I_{Cp} + I_{Cn} \tag{9.2}$$

where I_{Cp} is the hole component of the collector current and I_{Cn} is the electron component of the collector current. The total base current I_B is the difference between the total emitter current I_E and the total collector current I_C:

$$I_B = I_E - I_C \tag{9.3}$$

Consider the holes injected by the emitter in the pnp BJT. Most of these holes reach the collector without recombining in the base. Call the fraction of emitter holes reaching the collector B, the base transport factor, such that:

$$B = I_{Cp}/I_{Ep} \tag{9.4}$$

The hole-injection efficiency of the emitter is the ratio of the hole component I_{Ep} of the emitter current to the total emitter current I_E. Call this the emitter injection efficiency γ:

$$\gamma = I_{Ep}/(I_{Ep} + I_{En}) \tag{9.5}$$

Two important circuit design parameters are the DC common-base current gain α and the DC common-emitter current gain β, defined as follows:

$$\alpha = I_C/I_E \tag{9.6}$$

$$\beta = I_C/I_B \tag{9.7}$$

The common-base current gain α is almost unity because the total collector current I_C is almost as large as the total emitter current I_E. Typical values lie in the range 0.95–0.998. α is also known as α_{dc} (because it is a DC parameter only, not AC) and by the h-parameter h_{FB}.

The common-emitter current gain β, on the other hand, is much larger because I_B is very small. Typical values of β for small-signal pnp BJTs are in the range 20–500. β is the h-parameter h_{FE}, commonly quoted by manufacturers in their catalogues. It's also known as β_{dc}. These values of β make the BJT extremely useful as an amplifier: a small change in I_B causes a significant change in I_C. β is therefore an important parameter. There are problems associated with β: its value varies from device to device, even among devices from the same manufactured batch. β also varies with temperature and I_C. For these reasons minimum or 'worst' values of β are often quoted. High junction temperatures produce the highest values. The size of β reaches a peak as I_C is increased from zero but decreases again after about 10 mA in silicon BJTs.

We can determine β in terms of α and vice versa. Start off by writing equation (9.3) rearranged for I_E:

$$I_E = I_B + I_C$$

Dividing throughout by I_C gives

$$\frac{I_E}{I_C} = \frac{I_B}{I_C} + \frac{I_C}{I_C} = \frac{I_B}{I_C} + 1$$

From equations (9.6) and (9.7) we can now write

$$\frac{1}{\alpha} = \frac{1}{\beta} + 1 = \frac{1+\beta}{\beta}$$

Therefore the common-base current gain α can be written

$$\alpha = \frac{\beta}{1+\beta} \tag{9.8}$$

Rearranging for β:

$$\beta = \alpha(1+b) = \alpha + \alpha\beta$$
$$\beta - \alpha\beta = \alpha$$
$$\beta(1-\alpha) = \alpha$$

Finally, we obtain a common expression for the common-emitter current gain β:

$$\beta = \frac{\alpha}{1-\alpha} \tag{9.9}$$

We can also determine α in terms of γ if we assume $I_{Cn} \ll I_{Cp}$. From equation (9.2) we know that the total collector current I_C is the sum of the hole component I_{Cp} and the electron component I_{Cn}:

$$I_C = I_{Cp} + I_{Cn}$$

SELF-ASSESSMENT QUESTION 9.5

Why is the assumption $I_{Cn} \ll I_{Cp}$ valid?

If $I_{Cn} \ll I_{Cp}$ then we can write

$$I_C \approx I_{Cp} \tag{9.10}$$

The base transport factor B is given by equation (9.4):

$$B = I_{Cp}/I_{Ep}$$

so can be rewritten as

$$B = I_C/I_{Ep} \tag{9.11}$$

We know that α can be written (equation (9.6))

$$\alpha = I_C/I_E$$

Hence substituting from equation (9.1) gives

$$\alpha = I_C/(I_{Ep} + I_{En})$$
$$= BI_{Ep}/(I_{Ep} + I_{En})$$

Therefore

$$\alpha = B\gamma \tag{9.12}$$

Hence common-base current gain α is the product of base transport factor B and emitter injection efficiency γ. This makes intuitive sense because B is the fraction of emitter holes reaching the collector and γ is the ratio of the hole component I_{Ep} of the emitter current to the total emitter current I_E.

EXAMPLE 9.1

A certain BJT has the following known parameters: $\beta = 100$ for a base current I_B of 20 μA. Find the total collector current I_C and the common-base current gain α. To determine I_C use equation (9.7):

$$\beta = I_C/I_B$$

Therefore

$$I_C = \beta I_B$$
$$= 100 \times 20 \times 10^{-6} \text{ A}$$
$$= 2\text{mA}$$

To determine α use equation (9.8):

$$\alpha = \frac{\beta}{1+\beta}$$
$$= \frac{100}{101} = 0.99$$

The total collector current I_C is 2 mA and the common-base current gain α is 0.99.

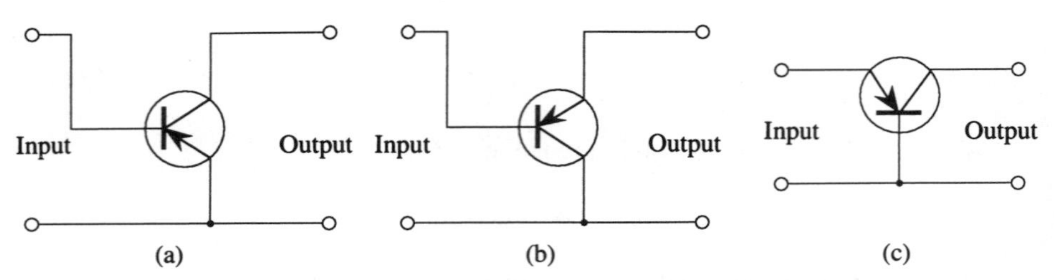

Figure 9.14 *Three common bias configurations for the pnp BJT: (a) common-emitter; (b) common-collector; (c) common-base.*

SELF-ASSESSMENT QUESTION 9.6

What value of voltage drop would you expect across a forward biased emitter junction in a silicon BJT? Explain your answer.

SELF-ASSESSMENT QUESTION 9.7

A silicon BJT has a β of 70 quoted by the manufacturer. What is the value of α for this transistor? What value of I_C would you expect for a base current of 20 μA?

9.4 Biasing the BJT for operation

There are several ways of connecting a bipolar junction transistor into a circuit. Each of these ways is called a bias configuration. The three most common are summarized in Figure 9.14.

Figure 9.15 illustrates the common-emitter configuration in more detail for both a pnp and an npn BJT. R_C is a collector resistor and R_B is the base resistor. The three terminal currents are shown. V_{CC} is the collector bias voltage and V_{BB} is the base bias voltage. V_{BB} forward biases the base–emitter junction and V_{CC} reverse biases the base–collector junction.

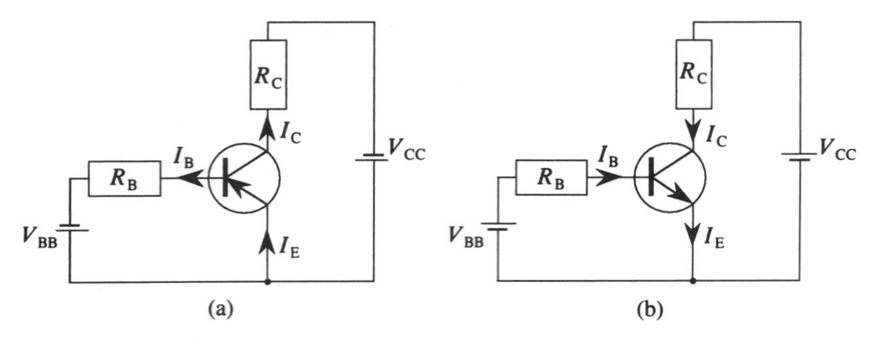

Figure 9.15 *Common-emitter bias circuits for the (a) pnp and (b) npn BJT.*

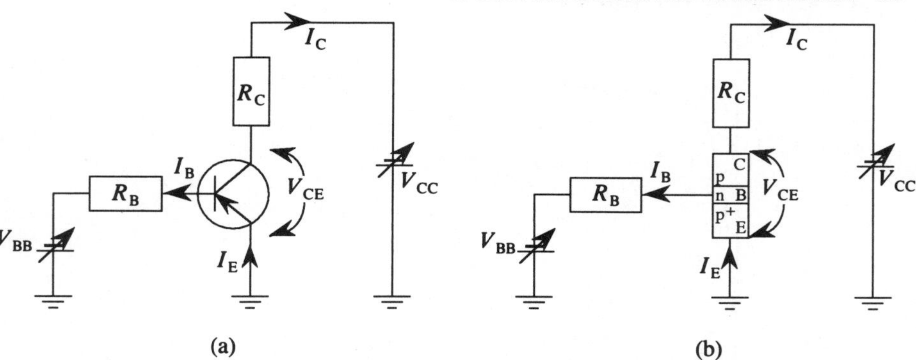

Figure 9.16 *Common-emitter circuit used to plot collector characteristics: (a) using the pnp BJT symbol; (b) replacing the symbol with a sketch of the pnp BJT.*

9.5 BJT characteristics

As with the FET, the operating conditions of a particular BJT are often summarized in the form of a family of collector characteristic curves. These curves are graphs of I_C versus V_{CE} for various values of I_B. The circuit used to determine these for a BJT is shown in Figure 9.16(a). Figure 9.16(b) is the same circuit, but the BJT symbol has been replaced by a sketch of the transistor; this has been done to help you. The bias voltages V_{BB} and V_{CC} are shown, along with the voltage drop across the transistor V_{CE}.

If both bias voltages V_{BB} and V_{CC} are set to zero there will be no currents flowing. If V_{BB} is then adjusted to give a non-zero value of I_B, a collector curve can be plotted for that I_B value. While V_{CC} is set to zero I_C will remain zero: this gives the first point on the characteristic. If V_{CC} is then increased, V_{CE} across the transistor will also increase and a current will flow from emitter to collector: I_C will no longer be zero and will increase quite sharply as V_{CE} is increased up to a certain value (Figure 9.17).

Figure 9.17 *Collector characteristic showing how I_C increases with increasing V_{CE} for low values of V_{CE}.*

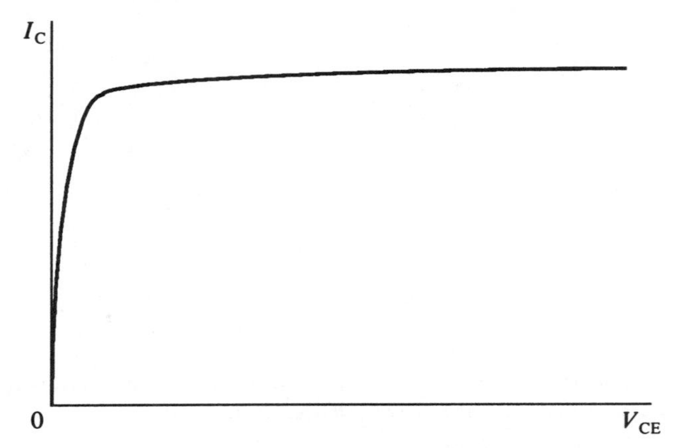

Figure 9.18 *A single collector characteristic curve for a BJT.*

If we assume the BJT is made of silicon, the emitter junction will 'turn-on' at about 0.7 V. The voltage dropped between the collector and the emitter will then be approximately the same and the base–collector junction will become reverse-biased. I_C will now reach its maximum value, found from equation (9.7):

$$I_C = \beta I_B$$

I_C levels off to an almost constant value as V_{CE} continues to increase (Figure 9.18). Note that I_C does continue to increase slightly as V_{CE} increases.

SELF-ASSESSMENT QUESTION 9.8

Why do you suppose I_C increases slightly as V_{CE} continues to increase?

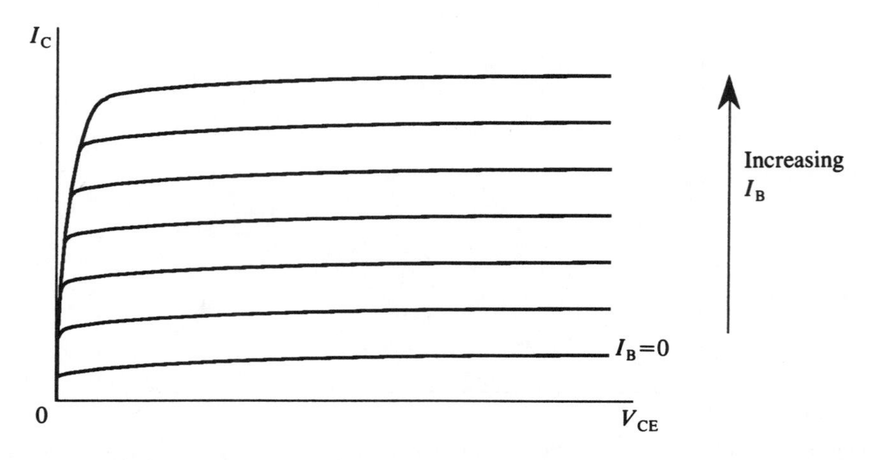

Figure 9.19 *A family of collector curves for a BJT.*

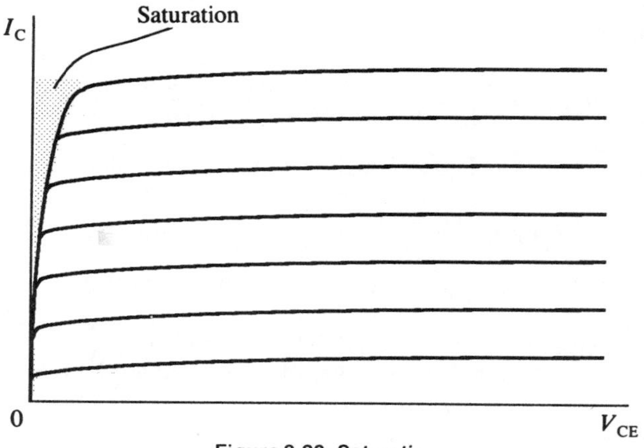

Figure 9.20 *Saturation.*

A family of characteristic curves will be obtained as the value of I_B is varied (Figure 9.19). When I_B is zero the BJT is said to be cut off. When I_B is zero, however, there will be a very small collector leakage current I_{CE0} consisting of thermally generated EHPs. This is very small and is often neglected. In cut-off both the emitter and the collector junctions are reverse-biased.

If the base–collector junction becomes forward biased I_C will not increase, even with increasing I_B. At this point we say the transistor is saturated. Saturation occurs at a V_{CE} value called $V_{CE(sat)}$. At saturation equation (9.7) is no longer valid. $V_{CE(sat)}$ is typically about 100 mV for silicon transistors. Figure 9.20 shows the saturation region on the family of collector curves.

EXAMPLE 9.2

Is the transistor of Figure 9.21 in saturation? Assume it's made of silicon, and assume $V_{CE(sat)}$ is negligible.

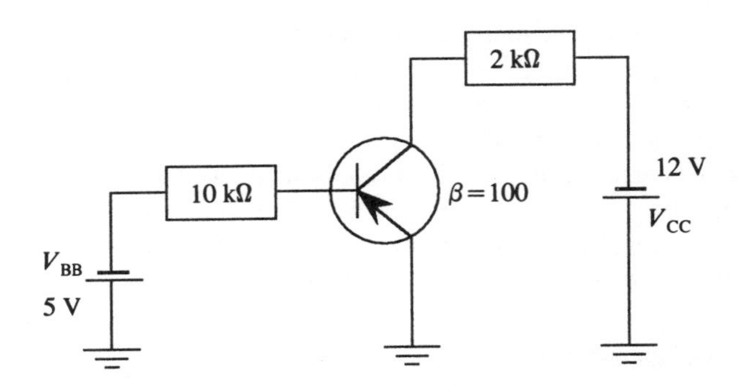

Figure 9.21 *Figure for Example 9.2.*

First determine $I_{C(sat)}$:

$$I_{C(sat)} = \frac{V_{CC} - V_{CE(sat)}}{R_C} \approx \frac{V_{CC}}{R_C} = \frac{12}{2 \times 10^3} = 6 \text{ mA}$$

Now find out if I_B is large enough to produce this value of $I_{C(sat)}$.

$$I_B = \frac{V_{BB} - V_{BE}}{R_B} = \frac{5 - 0.7}{10 \times 10^3} = 0.43 \text{ mA}$$

$$I_C = \beta I_B = 100 \times 0.43 \times 10^{-3} = 43 \text{ mA}$$

Hence the base current is able to produce I_C greater than $I_{C(sat)}$. The transistor is therefore in saturation. The I_C value of 43 mA cannot be reached.

SELF-ASSESSMENT QUESTION 9.9

What could you do to take the transistor of Figure 9.21 out of saturation?

9.6 Summary

The bipolar junction transistor is a three-terminal device which operates by injecting (diffusing) majority carriers from a heavily doped emitter region across a forward biased p–n junction called the emitter junction to the base. This injects a large number of carriers into the base region. There is also a p–n junction between the base and the collector which is reverse biased; this junction is called the collector junction. The reverse-biased collector junction causes the large number of minority carriers in the base to be swept across into the collector region, so forming a current. These transistors are designed to maximize the number of carriers leaving the emitter and entering the collector.

The collector junction becomes reverse biased when the voltage applied across the transistor V_{CE} is large enough to turn the base–emitter junction 'on' (0.7 V for a silicon BJT). When the transistor is switched on the collector current I_C increases with increasing base current I_B. Collector characteristics are plotted of collector current I_C versus the applied voltage across the collector and emitter of the transistor, V_{CE}, for various values of I_B.

Several BJT parameters have been defined, including the three terminal currents at the emitter, base and collector (I_E, I_B, I_C). The common-base current gain and the common-emitter current gain have been defined, and an equation relating these has been derived.

Bias configurations have been described briefly and a common-emitter circuit used for plotting collector characteristics has been explained.

9.7 Tutorial questions

9.1 How can recombination be minimized in the base region of an npn BJT?

9.2 What are the current components in the npn BJT? Which of them can be neglected?

9.3 A pnp BJT has $I_{Ep} = 1$ mA, $I_{En} = 0.01$ mA, $I_{Cp} = 0.88$ mA and $I_{Cn} = 0.1$ mA. Calculate the base transport factor, emitter injection efficiency, common-base current gain and base current.

9.4 A certain npn BJT has the following known parameters: $\beta = 240$ for a base current I_B of 10 μA. Find the total collector current I_C and the common-base current gain α.

9.5 A certain pnp BJT has the following known parameters: $\alpha = 0.998$ for a collector current I_C of 200 mA. Find the common-emitter current gain β, the base current I_B and the total emitter current I_E.

9.6 Sketch a circuit diagram to show how you might determine the common-emitter forward current gain β of an npn BJT.

9.7 Figure 9.22 shows a BJT connected in a circuit. Is the transistor npn or pnp? Is the transistor in saturation? Assume the transistor is made of silicon and state any other assumptions you make.

9.8 If the transistor of Figure 9.22 were made of germanium, what difference would it make, if any, to the saturation state of the transistor? Explain your answer.

9.8 Suggested further reading

Boylestad, R. and Nashelsky, L., *Electronic Devices and Circuit Theory*, 5th edn, Prentice Hall, 1992, pp. 108–129.

Ritchie, G. J., *h*-parameter modelling of BJTs, in: *Transistor Circuit Techniques: Discrete and Integrated*, 2nd edn, Van Nostrand Reinhold, 1987, pp. 161–164.

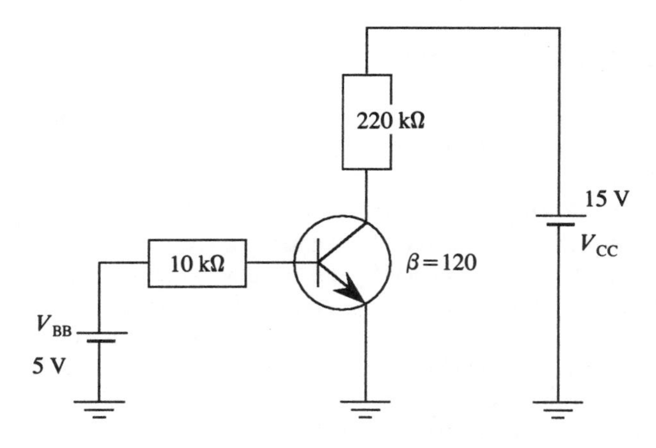

Figure 9.22 *Figure for tutorial questions 9.7 and 9.8.*

CHAPTER 10

Novel electronic materials

Aims and objectives

Technology is materials driven. Without materials scientists, we wouldn't have the silicon chip, helicopters, non-stick saucepans, lasers, ... The list is endless. Very often, as in the case of the helicopter (predicted by Leonardo da Vinci) and the laser (predicted by Albert Einstein), a product is envisaged and designed on paper. The paper design can only become a reality when the material is available to build the product. Einstein, for example, wrote a set of equations describing laser action but the first laser wasn't invented until 1960, five years after his death. As for the helicopter, it had to wait four hundred years until suitable alloys could be produced. Sometimes materials are discovered and researched without knowing what the eventual spin-offs will be: fullerenes are a good example of this. As the twenty-first century gets under way, will we be using flexible plastic sheets instead of the rigid and bulky computers used at the end of the 1990s? By 2010 will electronic circuits be organic, made from long-chain polymers such as plastics? The aims of this chapter, therefore, are to give a brief description of some of the novel materials which are currently being researched, to let you know some of the exciting recent developments, and to alert you to possible novel applications in the future.

10.1 Novel semiconductors

Semiconductor-device engineers are always searching for better and better semiconducting materials. In the first half of the twentieth century electronic circuits used valves — glass tubes containing gases — which were responsible for the electronic processing. Then Bardeen, Brattain and Shockley invented the transistor in 1948, and since then the second half of the twentieth century has seen massive changes in technology. Much of this upheaval has been the result of the search for materials with higher mobilities, lower costs, easier processing methods, and so on. The most commonly used semiconductor is silicon. This is because it's inexpensive (it's one of the most abundant elements on Earth) and because a massive financial investment has been made in designing factory-processing methods which are reliable and enable the mass production of silicon of very high quality. For a time gallium arsenide and the other III–V semiconductor compounds looked as though they'd knock silicon from its prime place in the semiconductor stakes, but as the twenty-first century arrives gallium arsenide and related compounds have made an impact only in the field of optoelectronic and microwave devices.

New forms of silicon are being investigated in the search for flat-screen display devices which would remove the need for the electron-gun tubes which are currently used in computer monitors and television sets, for example. A chemical method has been developed for the production of artificial diamond in the form of layers such that diamond can now be fabricated into semiconductor devices using planar technology. Another form of carbon that has been in the news is the fullerene: a molecule of carbon atoms in the form of a sphere. Polymer layers are now grown which have conducting or semiconducting properties, making it possible to produce organic electronic devices. The structure of electronic devices is becoming more and more complex as the dimensions of the layers become smaller and smaller. Nanotechnology is being used to produce planar devices which are only nanometres across.

10.1.1 **Heterojunction devices**

Many novel devices use semiconducting materials in the form of heterostructures. A p–n junction between two layers of the same material is called a homojunction, and the rectifying energy barrier in the junction is brought about by the difference in the fermi-level positions either side of the junction. Rectification can also be brought about by dissimilarities in bandgap: a p–n junction between two materials of different bandgap is called a heterojunction. Heterojunction devices can even be made such that each side of the junction is of the same conductivity type (i.e. both n-type or both p-type) — this is an *isotype* heterojunction. Heterojunctions in which one side is n-type and the other p-type are called *anisotype*. Figure 10.1 illustrates the differences between homojunctions and heterojunctions.

Lattice matching is important in heterojunctions because the material needs to be as single-crystal-like as possible to maximize the free movement of electrons. For example, from Table 3.2 you can see that aluminium arsenide AlAs and gallium arsenide GaAs have very similer lattice constants, 5.63 Å and 5.64 Å respectively. If these are alloyed to give AlGaAs the alloy retains the lattice constant of about 5.64 Å so that a layer of AlGaAs can be grown on to a layer of GaAs without distorting the lattice. From Table 6.2 you can see that GaAs has a bandgap of 1.43 eV at room temperature, whereas $Al_{0.3}Ga_{0.7}As$ has a bandgap of 1.87 eV at room temperature. This difference in bandgap can be used to produce discontinuities in the band edges ΔE_c and ΔE_v (Figure 10.2), such that the difference in bandgap is equal to the sum of the discontinuities. For the AlGaAs/GaAs system of Figure 10.2,

$$1.87 - 1.43 = 0.44 \text{ eV}$$

and

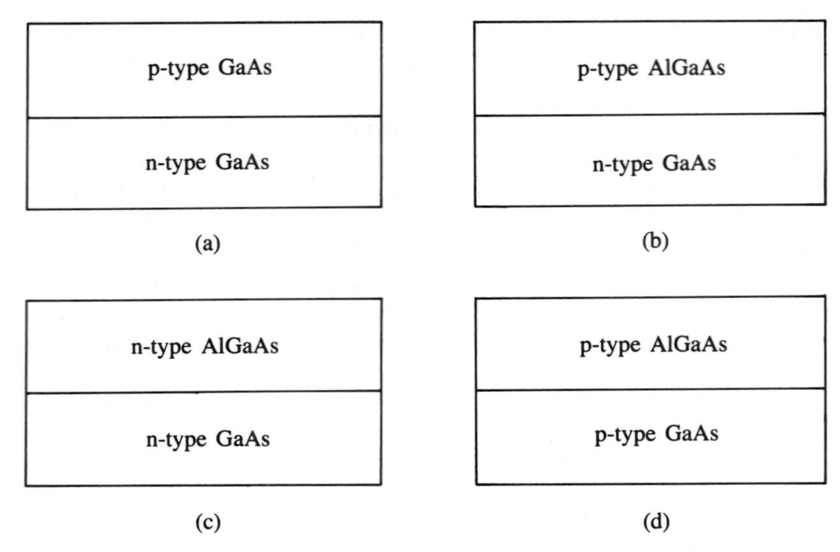

Figure 10.1 *Four types of semiconductor junction: (a) homojunction; (b) anisotype heterojunction; (c) n-type isotype heterojunction; (d) p-type isotype heterojunction.*

$$\Delta E_c = 0.29\,\text{eV}$$
$$\Delta E_v = 0.15\,\text{eV}$$

The values of the discontinuities ΔE_c and ΔE_v can be found by taking capacitance–voltage measurements (Section 5.5.2).

SELF-ASSESSMENT QUESTION 10.1
Sketch the energy-band diagram for an anisotype heterojunction of n-type AlGaAs on p-type GaAs. Assume the following: E_g for AlGaAs = 1.87 eV; E_g for GaAs = 1.43 eV; ΔE_c = 0.29 eV; ΔE_v = 0.15 eV.

Heterostructures are used to obtain faster device speeds so that they can produce switching at very high frequencies, in the gigahertz range. They are also invaluable for laser diodes as the bandgap difference can be used to physically confine the carriers and the laser light to enable efficient operation. Sometimes lasers are designed with two heterojunctions, called double heterojunction lasers (Figure 10.3).

10.1.2 Amorphous silicon and polysilicon

An amorphous semiconductor is composed of a disorderly array of tiny crystals. It is not, therefore, a single-crystal material. The most commonly used amorphous semiconductor is amorphous silicon, or α-silicon. It is thought that the tetrahedral nature of the silicon bonds in the single-crystal version (Figure 3.18(a)) is retained, but

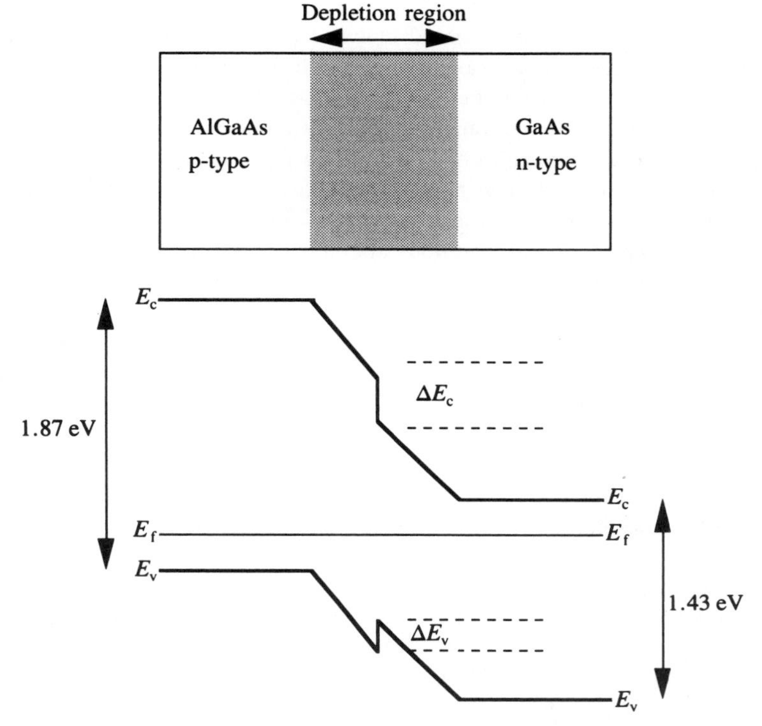

Figure 10.2 *Energy-band diagram for an anisotype heterojunction between p-type AlGaAs and n-type GaAs.*

that the angles and bond lengths are distorted. Also, some of the bonds are left 'dangling', meaning that a partner atom for the bond is not always found. These dangling bonds give rise to a lot of energy states within the bandgap. The bandgap in α-silicon is direct and is about 1.6 eV, significantly larger than that in single-crystal silicon. This makes it particularly useful for use as a solar-cell material.

SELF-ASSESSMENT QUESTION 10.2

What wavelength is 1.6 eV equivalent to?

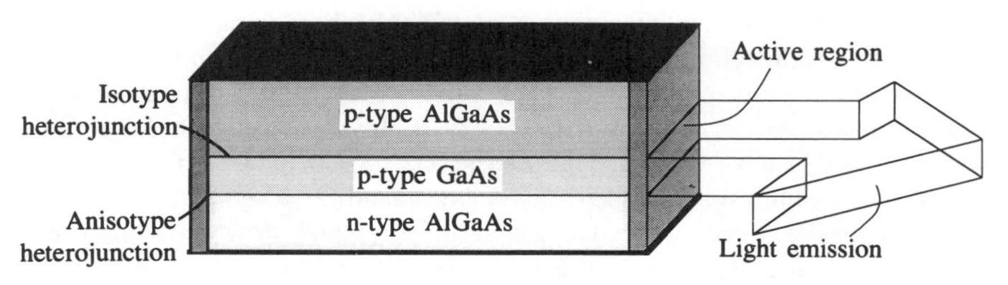

Figure 10.3 *A simple double heterojunction laser diode.*

Amorphous silicon is also a good material for the production of liquid-crystal displays, so is ideal for the production of small-display devices such as miniature televisions and palm-top computers and notebooks. Amorphous-silicon technology is good enough to fabricate the transistors that are embedded in screens to control the light level, but the same technology cannot be used to fabricate the drive circuits needed; these drive circuits have to be manufactured separately and then connected to the display. This is a disadvantage of using α-silicon for display devices.

Polysilicon is made from amorphous silicon which is annealed. The polysilicon has to be deposited onto a substrate to give it rigidity so that it can be annealed satisfactorily. The high-temperature annealing process requires expensive quartz substrates to be used, leading to great expense for the manufacturers. As a consequence of this a lot of research is being done into a low-temperature annealing method, which can utilize ordinary glass substrates.

RESEARCH 10.1

Find out the melting points of quartz and ordinary glass (such as crown glass), and note the temperatures in your notebook. Don't forget to note down the references.

One advantage that polysilicon has over amorphous silicon is that the transistors and the drive circuitry can be produced on the same chip. This means the pixel density can be higher than that found in amorphous-silicon devices, as the transistor–drive circuit unit will be considerably smaller. Display devices made of polysilicon will therefore have a higher resolution than α-silicon devices. Large-format polysilicon displays (e.g. large enough for televisions) are not expected until the 21st century.

10.1.3 **Semiconducting diamond and fullerenes**

Solid carbon exists in two commonly known forms: graphite and diamond. Graphite is a soft, flaky substance with a plate-like structure which sheds layers of atoms so readily that we use it in 'lead' pencils. Diamond is completely different: it is a hard, strong material which can be polished to make gemstones. It is the hardest naturally occurring substance known to Humankind — it will cut anything softer than itself.

In recent years a third form of carbon has been discovered: the buckminsterfullerene, or fullerene for short. These are also called 'Bucky balls' because of their curious spherical shape. One 'ball' is made up of one molecule composed entirely of about 60 carbon atoms. They've been found in clay and in soot deposits, and prehistoric rocks. They're made entirely of carbon atoms arranged in a hollow patchwork sphere, the patches of which are hexagons and pentagons with a carbon atom at each point. The shape is exactly like a soccer ball. At the time of writing, nobody knows what fullerenes can be used for. Could we use them to transport atoms or ions held inside the sphere? If so, how could we open up the spheres to put something inside? Could we find a mechanism of opening them up to produce long chains? Fullerenes are cheap and easy to produce by burning carbon, so a lot of scientists will be trying to find answers to these questions.

RESEARCH 10.2

Find a diagram of the crystal structure of graphite and compare it with the diamond crystal illustrated in Figure 3.18.

Until recent years diamond was mined at great expense or was produced by subjecting graphite to extremely high pressures and temperatures, causing it to re-form as diamond. It has important commercial applications as a cutting tool, so a cheaper means of production has been sought over the years. These tools are coated with very many tiny diamonds, so they're very expensive. Cheaper versions of cutting tools are widely available, but they're not so efficient. For most of the twentieth century scientists have been trying to produce synthetic diamond from hydrocarbon gases (typically methane CH_4). In the 1950s Russian scientists found that diamond layers could be deposited onto a substrate by a process called chemical vapour deposition (CVD) in which a mixture of gases containing carbon atoms was allowed to solidify on a diamond-powder substrate. Since then it has been possible to grow diamond layers by CVD on non-diamond substrates.

Chemical vapour deposition is a commonly used process in the semiconductor industry, so diamond was suddenly brought to the attention of semiconductor-device engineers as a potentially useful electronic material. It had been known for a long time that diamond has a very wide bandgap, about 6 eV, but it was also known that diamond has a very high mobility. As the two most desirable electrical properties of a semiconductor are that they have lots of carriers and that the carriers have high mobility, interest in diamond grew. After all, if it were possible to dope the diamond sufficiently so that many carriers occupied the bandgap, couldn't sufficiently high levels of conduction be achieved? Work has been carried out since the 1980s to make diamond a suitable high-mobility semiconductor for use in the electronics industry. The method of production has been improved and semiconducting layers of diamond have been produced.

There are many possible future applications of synthetic diamond. Its semiconducting properties can be used in transistors, optoelectronic switches, radiation detectors and lasers.

10.2 **Superconductors**

In 1911 Kamerling Onnes discovered superconductivity, and was awarded the Nobel Prize in 1913 for doing so. He did this by cooling mercury to very low temperatures. Lord Kelvin had predicted that metals cooled to very low temperatures would have such a high electrical resistance that electrons travelling through the metal would be enormously slowed down. In fact, Onnes observed the opposite effect: the electrons seemed to be travelling through the metal without any electrical resistance at all. He realized that he had made an important discovery because he knew that electrical resistance was responsible for the limits on the usefulness of electrical materials. He suggested that high-field magnets would be built one day, and in fact this is now one of the major uses of superconducting materials.

RESEARCH 10.3
What is magnetic resonance imaging, and why is its application important in medicine? See if you can find the answers to these questions.

Until 1986 superconducting materials had to be cooled to very low temperatures, less than 30 K, before superconductivity could be observed. Liquid helium was used to do this because it cooled the superconductor to a few kelvin. This wasn't very convenient and was also very expensive, and so people continued to search for ways of increasing the temperature at which superconductors could be used. After all, if electronic devices or electrical machines were to take advantage of superconducting materials, they'd have to use these materials at normal temperatures (say from 0°C up to 50°C). A major aim of researchers has been to find a material that would superconduct at room temperature, about 300 K. Many conventional electrical devices actually operate at temperatures higher than room temperature — if you put your hand near a small transformer, for example, you'll feel the warmth radiating from it. Inside the transformer it could be warmer than 350 K; this is because of the work the electrons expend in travelling around the resistive material — the work done produces heat. All current-carrying wires and devices likewise produce heat because of the work done in overcoming their resistance.

RESEARCH 10.4
What is the boiling point of helium?

10.2.1 **Properties of superconductors**

Superconducting properties are destroyed by too much current, too great a magnetic field, and too high a temperature. Consequently superconductors have a critical current density, a critical magnetic field, and a critical temperature.

The critical temperature is the temperature below which materials superconduct. Prior to 1986 it was thought that this critical temperature could not be higher than about 30 K. Muller and Bednorz, scientists at the IBM Laboratories in Zurich, won the Nobel Prize in 1986 when they observed superconductivity in cuprate materials at 30 K in a lanthanum cuprate. Then in 1987 Chu observed superconductivity at 90 K by replacing the lanthanum by yttrium. This was especially important because this was the first time superconductivity had been reported at a temperature higher than the boiling point of nitrogen (77 K). In the years since then superconductivity has been observed at higher and higher temperatures in a variety of cuprates. The critical temperature is now thought to be higher than 130 K.

The theory of superconductivity developed by Bardeen, Cooper and Schreiffer in 1957 (known as the BCS theory) postulates the existence of pairs of electrons, called Cooper pairs, which carry the super current. It is thought that these Cooper pairs break up when the temperature, current density or magnetic field is too high. However, the theory describing high-temperature superconductivity is still young, and there are lots of anomalies which cannot yet be explained.

Current research is focused on high-temperature superconducting materials which are ceramic, including $YBa_2Cu_3O_7$ (called YBCO).

10.3 **Conducting polymers**

Polymers consist of long organic molecules, the 'backbones' of which are made of chains of carbon atoms (Figure 10.4). The long-chain structure enables conduction by *charge hopping* along the chain. If the individual molecules are bonded together to form a fibre, there will be more opportunities for conduction. These fibres can also be produced in the form of a thin film.

Figure 10.4 represents a long-chain organic molecule such as polyacetylene. Carbon has a valency of four, so each carbon atom has four valence electrons which it can share with its neighbouring carbon atoms. It does this by forming *single* and *double* bonds alternately along the chain. If a dopant atom such as iodine is then used to dope the chain, the iodine will take an electron from one of the double bonds to form a negative ion, leaving a positive hole on the carbon chain (Figure 10.5). A neighbouring electron in a double bond will hop into the hole, and so on, to produce hole conduction.

Polyacetylene and polyethylene exhibit such hopping conduction, and they even have an energy-band structure similar to that of crystalline semiconductors. The bandgap in polyacetylene is about 1.4 eV. These polymeric films can also be doped in the same way as a crystalline semiconductor. For example, polyacetylene can be doped with iodine to produce dopant states in the bandgap. Dopants are commonly chosen from groups IA (for p-type) and VIIB (for n-type) of the periodic table.

There are many more free carriers in polymers than there are in crystalline semiconductors such as gallium arsenide and low resistivities (about $10^{-2}\Omega$ cm) can

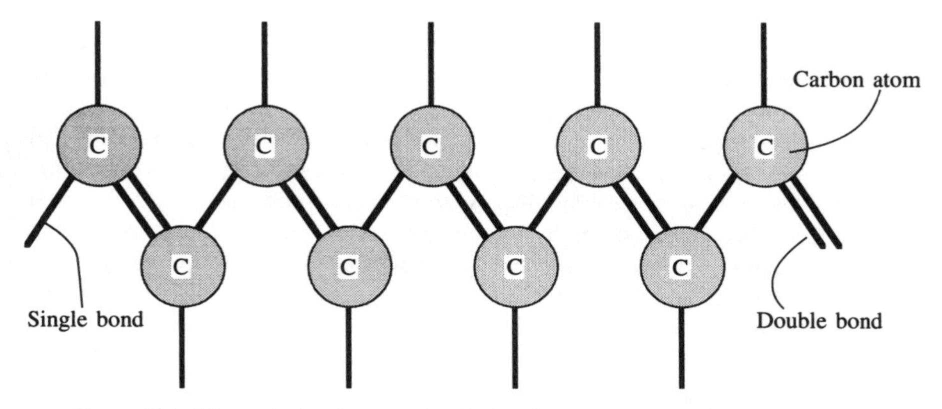

Figure 10.4 *A long-chain polymer molecule showing alternate single and double bonds.*

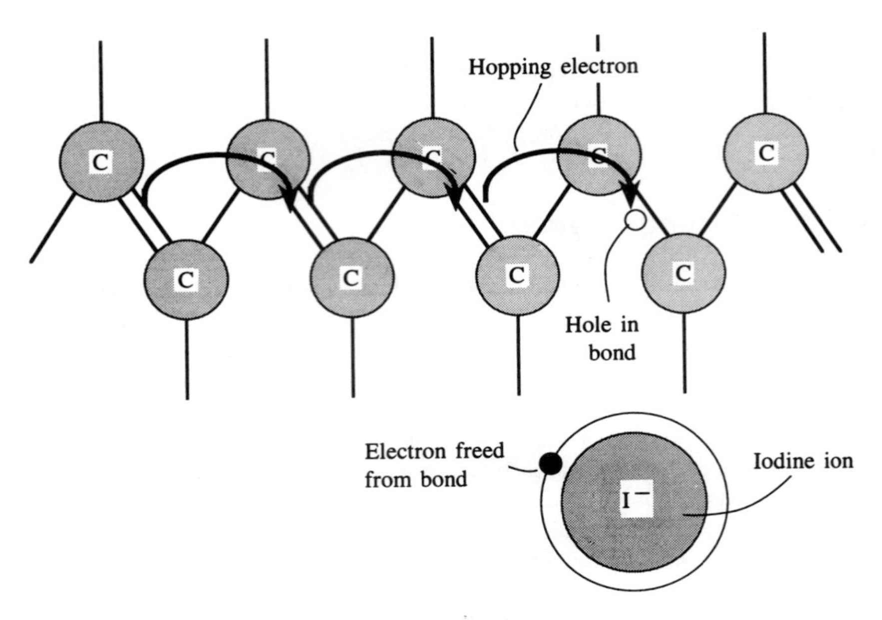

Figure 10.5 *Hopping conduction in a conducting polymer.*

be obtained by doping, but unfortunately mobilities are low also. Mobility can be increased by straightening out the molecules into a thin sheet, thereby aligning them. Using this method, conductivities have been achieved which are as high as in some metals.

10.4 **Summary**

Brief descriptions of some novel materials and structures have been given: semiconductor heterojunction devices, amorphous silicon and polysilicon, semiconducting diamond, fullerenes, superconductors and conducting polymers. Heterojunction devices are being made because of their high switching speeds and optoelectronic efficiency. New types of silicon are being investigated as part of the search for larger and larger flat display screens. Diamond is of interest as a semiconductor because of its high mobility. Fullerenes have been discovered but haven't yet found a use, unlike superconductors which have a very promising future for a range of electrical applications from power generation to medicine. Polymers have been made which behave like semiconductors. These are just a small selection of the many developments there have been in electronic-materials technology over recent years. Readers are encouraged to read widely in newspapers, magazines and journals to discover more information about novel materials.

10.5 Suggested further reading

Alford, N. McN., Applications of high temperature superconductors. *Chemistry and Industry*, 2 November 1992, pp. 809–812.

Chinnock, C., Polysilicon: the path to better displays?, *Byte*, November 1994, p. 34.

Diamond to order, *New Scientist*, 25 February, 22, 1995.

Fox, K.C., The electric plastics show, *New Scientist*, **141** (1915), 33–37, 1994.

Grovenor, C.R.M., *Microelectronic Materials*, Adam Hilger, 1989, pp. 43–47.

Guinier, A. and Jullien, R., *The Solid State: from Superconductors to Superalloys*, Oxford University Press, 1989, pp. 126–134.

Mandell, M., Diamond helps, *New Scientist*, **129**, (1758), 27, 1991.

New flat-screen technology, *Computer Shopper*, June, 411, 1995.

Streetman, B.G., *Solid State Electronic Devices*, 4th edn, Prentice Hall, 1995, pp. 281–283, 392–397.

Answers to self-assessment questions

Chapter 1

1.1 Your answer should be 4.85×10^{-7} m or 0.485 μm. It's important that you know 1 μm = 1×10^{-6} m as microns (i.e. micrometres) are commonly used in solid-state electronics. If you got the wrong answer you may have got p and j the wrong way round in the equation (if you did, was your answer negative?), or you may have forgotten to invert $1/\lambda_{pj}$.

1.2 What did you get for your answer? It should have been 5.3×10^{-11} m. Make sure you got the units correct — if you gave your answer in metres it should have been 5.3×10^{-11} m, but if you gave your answer in nanometres, for example, it should have been 0.053 nm. You could also use angstrom units, Å, in which case your answer should have been 0.53 Å. In picometres it should have been 53 pm. You'll discover the significance of the -2.18×10^{-18} J later on.

1.3 Well, what did you get for your answers? The energy of a 1 μm photon is 2×10^{-19} J, and its frequency is 3×10^{14} Hz. If your answers don't agree, it could be because you don't know what a micron (μm) is, or because you've misused your calculator. Watch out for calculator errors — students continually lose marks in exams and coursework because they don't know how to use a calculator. You should always have an idea of the answer before you use your calculator; I suggest you estimate your answer either in your head or in a brief aside to your notes before you pick up your calculator.

1.4 Substitute values of the constants into $e^4 m_o / 8\epsilon_o^2 h^3 c$ to give:

$$\frac{e^4 m_o}{8\epsilon_o^2 h^3 c} = \frac{(1.602 \times 10^{-19})^4 \times 9.109 \times 10^{-31}}{8 \times (8.854 \times 10^{-12})^2 \times (6.626 \times 10^{-34})^3 \times 2.998 \times 10^8}\ \mathrm{m}^{-1}$$

$$= \frac{1.602^4 \times 9.109}{8 \times 8.854^2 \times 6.626^3 \times 2.998} \times \frac{(10^{-19})^4 \times 10^{-31}}{(10^{-12})^2 \times (10^{-34})^3 \times 10^8}$$

$$= \frac{6.586 \times 9.109}{8 \times 78.393 \times 290.907 \times 2.998} \times \frac{10^{-76} \times 10^{-31}}{(10^{-24}) \times 10^{-102} \times 10^8}$$

$$= \frac{59.992}{546956.858} \times \frac{10^{-107}}{10^{-118}}$$

$$= 0.0001097 \times 10^{11}$$

$$= 10\,968\,324$$

$$= 1.097 \times 10^{7}\ \mathrm{m}^{-1}$$

This is the Rydberg constant. There is a quicker way of finding the answer than the laborious method I've shown here. You could perform the calculation in one elaborate step on an electronic calculator, but in my experience students often make errors when doing so. I suggest you estimate the answer as follows:

$$\frac{e^4 m_o}{8\epsilon_o^2 h^3 c} \approx \frac{(2 \times 10^{-19})^4 \times 10 \times 10^{-31}}{8 \times (9 \times 10^{-12})^2 \times (7 \times 10^{-34})^3 \times 3 \times 10^8}$$

$$\approx \frac{16 \times 10}{8 \times 80 \times 350 \times 3} \times \frac{(10^{-19})^4 \times 10^{-31}}{(10^{-12})^2 \times (10^{-34})^3 \times 10^8}$$

$$\approx \frac{2}{8 \times 1000} \times \frac{10^{-76} \times 10^{-31}}{10^{-24} \times 10^{-102} \times 10^8}$$

$$\approx 0.1 \times 10^{-3} \times 10^{11}$$

$$\approx 1 \times 10^{7}\ \mathrm{m}^{-1}$$

This time I've reached approximately the expected answer by reducing the number of significant figures in each part of the calculation and without needing to use a calculator. With practice you should be able to perform estimations like this very quickly and with sufficient accuracy.

1.5 You should have found that the ionization energy of hydrogen is -2.18×10^{-18} J, or -13.6 eV. Make sure you got the correct units. Did you get an answer of zero? If so, did you follow earlier advice about estimating your answer before picking up your calculator and, therefore, did you expect a non-zero answer? If you did get zero you should have realized immediately that you'd made a mistake. Do the calculation again, and this time deal with the exponents in the equation separately, preferably without using your calculator! Zero answers arise when the numbers you're using are outside the range of numbers your calculator can cope with.

Chapter 2

2.1 The four quantum numbers are the principal quantum number n, the orbital quantum number l, the magnetic quantum number m and the spin quantum number s.

2.2 C 6, O 8, N 7, Si 14, Ge 32, Fe 26, Ne 10, Cl 17, Hg 80, Ag 47, Y 39, K 19, Pb 46, W 74.

2.3 According to the Periodic Table there are four electrons in beryllium. The first three take the configurations of those in lithium. The fourth has values $n=2, l=0, m=0, s=-\frac{1}{2}$.

2.4 This one's a bit trickier than beryllium because a new value of the orbital quantum number l is required for $n=2$. Boron has atomic number $Z=5$, so it has five electrons. The configurations of the first four electrons are the same as those in beryllium. The fifth electron has the following quantum numbers: $n=2, l=1, m=-1, s=+\frac{1}{2}$.

2.5 If you apply the rule which determines how many different m values an electron can take for a particular value of l, and if you remember that there are two spin states, you should find that there can be 2 s electrons, 6 p electrons, 10 d electrons and 14 f electrons.

2.6 The carbon ion C^{4+} arises when a neutral carbon atom has four electrons removed, so its electronic configuration is $1s^22s^2$. Likewise, the electronic configuration of Na+ is $1s^22s^22p^6$. The negative ions O^{2-} and $C1^-$ have gained electrons, so their configurations are $1s^22s^22p^6$ and $1s^22s^22p^63s^23p^6$ respectively.

2.7 There are 26 electrons in a neutral atom of iron.

2.8 Use the $2n^2$ rule: the maximum number of electrons that can go into the $n=3$ shell is eighteen.

2.9 There is a maximum of ten d electrons in a shell because d electrons are those electrons for which the orbital quantum number $l=2$, hence there are five values of magnetic quantum number m that a d electron can have $(-2, -1, 0, +1, +2)$ and each of these can have a spin quantum number s of $-\frac{1}{2}$ or $+\frac{1}{2}$.

Chapter 3

3.1 Free electrons wander around the lattice of the material by moving from empty state to empty state, from one atom to another. If an electric field is applied to the material the free electrons wander towards the positive pole, thereby forming a current.

3.2 Silicon dioxide, SiO_2, because its bandgap is very wide. This means it's very unlikely that electrons will get enough energy to cross the bandgap and thereby become available for conduction.

3.3 There are four atoms contained within one FCC unit cell: one-eighth at each of the eight corners, and half at the centre of each of the six faces.

3.4 (010) plane.

3.5 $(\bar{1}00)$ plane.

3.6 $(\bar{1}11)$ plane.

3.7 (203) plane.

Chapter 4

4.1 $1 \times 10^{12}\,\mathrm{cm^{-3}}$, because the density of electrons in the conduction band must equal the density of holes in the valence band.

4.2 Yes, the density would change. The hole density will still equal the electron density.

4.3 The activation energy is given by $E_g - E_d$, i.e. $1.1 - 0.9 = 0.2\,\mathrm{eV}$.

4.4 The activation energy is given by E_a, i.e. 0.6 eV.

4.5 The voltage required is 4.3 V.

4.6 The energy required is 1.4 eV.

4.7 The thermal velocity of an electron at room temperature is about $1 \times 10^5\,\mathrm{m\,s^{-1}}$.

4.8 The drift velocity would be $0.85\,\mathrm{m\,s^{-1}}$. This is about 10^5 times smaller than the thermal velocity $1\times10^5\,\mathrm{m\,s^{-1}}$.

4.9 The mean free time between collisions τ would be about 5×10^{-12} s, or 5 ps.

4.10 The conductivity is $5\,\mathrm{m\Omega^{-1}\,cm^{-1}}$.

4.11 The electron drift mobility is $0.78\,\mathrm{m^2\,V^{-1}\,s^{-1}}$.

4.12 The electron drift mobility is $7800\,\mathrm{cm^2\,V^{-1}\,s^{-1}}$.

4.13 The conductivity is $0.01\,\Omega^{-1}\,\mathrm{cm^{-1}}$.

4.14 The resistivity is $0.05\,\Omega\,\mathrm{m}$ or $5\,\Omega\,\mathrm{cm}$.

4.15 The electron diffusion-current density $J_{n(diff.)}$ would be $3.5\times10^{-13}\ \mathrm{A\,cm^{-2}}$ or $0.35\ \mathrm{pA\,cm^{-2}}$.

4.16 According to the Einstein relationship, the electron diffusion coefficient should be $64.7\ \mathrm{cm^2\,s^{-1}}$.

4.17 The energy of a 1300 nm photon is 0.96 eV, the energy of a 644 nm photon is 1.93 eV and the energy of a 480 nm photon is 2.59 eV. Therefore the 644 nm and 480 nm photons have energies greater than the bandgap energy, so would be absorbed by the gallium arsenide. The sample would be transparent to the 1300 nm photon.

4.18 The 1300 nm photon is infra-red, the 644 nm photon is red and the 480 nm photon is blue. The red and blue wavelengths mentioned here are very close to two standard light sources which are used for calibration purposes: the cadmium red line (643.84696 nm) and the cadmium blue line (479.991 nm).

Chapter 5

5.1 This is because the energy-band diagram is drawn from the point of view of an electron. Remember the electron has a negative charge so it 'climbs' in the opposite direction to holes, which are positive.

5.2 Drift current is small because it is made up of minority carriers, and there aren't many of them.

5.3 The barrier height is reduced, so more majority carriers are able to cross it at normal temperatures. Drift stays the same as in the equilibrium case because the minority carriers don't have to 'climb' a barrier, therefore they are just as likely to cross as in the equilibrium case.

5.4 You should expect that they would be similar. It's unlikely that they would be identical because they're defined differently.

5.5 The barrier height is increased, so fewer majority carriers are able to cross it at normal temperatures. Drift stays the same as in the equilibrium case because the minority carriers don't have to 'climb' a barrier, therefore they are just as likely to cross as in the equilibrium case and the forward-bias case. The number of drift carriers remains small.

5.6 The diode conducts in one direction only because the barrier height is reduced in forward bias, enabling more majority carriers to diffuse across the barrier.

5.7 The reverse current is made up of minority carriers which drift.

5.8 If $N_a \gg N_d$, then $1/N_d \gg 1/N_a$. This means $1/N_a$ in equation (5.26) can be neglected.

5.9 The depletion width in a pn$^+$ junction is given by

$$W \approx x_p = \left[\frac{2\epsilon_s(V_o - V)}{eN_a}\right]^{1/2}$$

because $N_a \ll N_d$, hence N_d in equation (5.26) can be neglected.

5.10 The slope is equal to $-2/e\epsilon_o\epsilon_r A^2 N_a$; all these quantities except N_a can be measured, so N_a can be found. The intercept is equal to $2V_o/e\epsilon_o\epsilon_r A^2 N_a$; if N_a is determined first the only unknown here is V_o, enabling it to be found.

Chapter 6

6.1 4.4×10^{-31} kg.

6.2 0.66 μm ($Al_{0.3}Ga_{0.7}As$) and 0.92 μm (InP). The former emission in $Al_{0.3}Ga_{0.7}As$ is visible, the wavelength being in the red part of the spectrum. The emission from InP is in the infra-red, therefore invisible to the human eye.

6.3 The probability of finding an electron at the conduction-band edge at room temperature is 0.40, or 40%. Use the Fermi–Dirac distribution function and substitute in the following values: $E_c — E_f$ is 0.01 eV and kT is 0.025 eV.

6.4 2.9×10^{25} m^{-3} or 2.9×10^{19} cm^{-3}.

6.5 $n_o = 1.2 \times 10^{25}$ m^{-3} or 1.2×10^{19} cm^{-3}.

6.6 The sample is n-type because the Fermi level is only 0.01 eV below the conduction-band edge. This sample is silicon so the bandgap at room temperature is 1.1 eV, hence E_c is much higher than the intrinsic Fermi level E_i which would be 0.55 eV below E_c.

6.7 1.5×10^{16} m^{-3}.

Chapter 7

7.1 The reverse current I_o is small and constant because it's made up of minority carriers, and there are not many of them.

7.2 The turn-on voltage arises because of the energy barrier (or contact potential V_o) between the two sides of the p–n junction. The forward diffusion current does not increase significantly until the applied bias is at least as large as the turn-on voltage because carriers diffusing from the p-side to the n-side have to acquire enough energy to climb the energy barrier before a significant amount of conduction can occur. They acquire enough energy only when the contact potential V_o (approximately equal to $V_{turn\text{-}on}$) is reached.

7.3 You'll need to estimate these answers by taking readings from the graph to determine the slopes of the dark and light reverse characteristics. $R_{reverse}$ under dark conditions is approximately 400 kΩ and the minimum $R_{reverse}$ under light conditions is approximately 100 kΩ. Don't forget the slope is dI/dV but the resistance is the inverse of this.

7.4 The long-wavelength cutoff will be given by $h\,c/e\,E_g$, therefore it is 900 nm for silicon.

7.5 You need to draw in the maximum-power rectangle as shown in Figure SAQ 7.5 to get an answer of approximately 0.8, as follows.

$$\text{Fill factor} = V_{max} I_{max} / I_{sc} V_{oc}$$

$$= 0.23 \times 70 \times 10^{-6} / 80 \times 10^{-6} \times 0.26$$

$$= 0.77$$

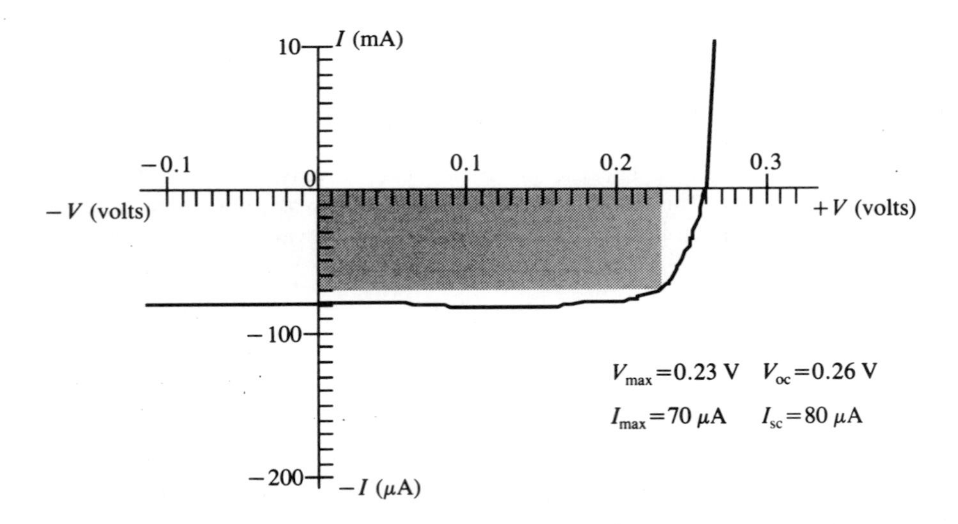

Figure SAQ 7.5 *The maximum-power rectangle for the solution of SAQ 7.5.*

7.6 The bandgap transition results in the emission of a photon of wavelength 0.89 μm. A radiative transition from the conduction-band edge to the donor site would produce a wavelength of 207 μm. The first wavelength is in the infra-red, whereas the second is very long (microwaves).

Chapter 8

8.1 The majority carriers are electrons.

8.2 In the p-channel JFET shown the drain current travels from the source to the drain, and the majority carriers (holes) also travel from the source to the drain.

8.3 The gate is highly doped relative to the channel because this produces a wide, controllable depletion region in the n-channel and not in the p-type gate material (the junction is p^+n).

8.4 The gate–drain junction is reverse biased because the p-type region (the gate) is negatively biased relative to the n-type region (the drain). The gate–source junction is forward biased because the p-type region (the gate) is positively biased relative to the n-type region (the source).

8.5 The channel resistance is given by Ohm's law, so $R_{\text{channel}} = V_{DS}/I_D = 6/12 = 0.5\,\Omega$.

8.6 The cut-off voltage $V_{GS(\text{off})}$ for the characteristics shown is -7 V.

8.7 Channel resistance when $V_{GS} = -1$ V is 100 Ω; channel resistance when $V_{GS} = -5$ V is 830 Ω; pinch-off voltage V_p is 7 V; steady-state drain current I_{DSS} at $V_{GS} = 0$ V is 70 mA; pinch-off voltage when $V_{GS} = -1$ V is 6 V; pinch-off voltage when $V_{GS} = -5$ V is 5 V; breakdown voltage when $V_{GS} = -1$ V is 25 V; breakdown voltage when $V_{GS} = -5$ V is 20 V.

8.8 V_{GS} should be biased such that the gate is positive and the source is negative, and V_{DS} should be biased such that the source is positive relative to the drain.

Chapter 9

9.1 Diffusion length is the average distance a carrier travels between generation and recombination. Lifetime is the average time a carrier is free between generation and recombination.

9.2 There will be four current components across each junction in the pnp BJT. They are: electron diffusion current, hole diffusion current, electron drift current and hole drift current.

9.3 There are fewer minority electrons in the collector than minority holes in the base because lots of holes are injected from the emitter into the base.

9.4 The hole and electron drift currents across the emitter junction (because it is forward biased) and the hole and electron diffusion currents across the collector junction (because it is reverse-biased).

9.5 The assumption $I_{Cn} \ll I_{Cp}$ is valid because many holes reach the collector from the emitter, and the number of electrons in the collector is small because they are the minority carriers there and will recombine.

9.6 When a silicon emitter junction is forward biased it will have a voltage drop V_{BE} of about 0.7 V because this is the turn-on voltage in a silicon p–n junction.

9.7 The value of α is 0.986. The value of I_C is 1.4 mA.

9.8 I_C increases slightly as V_{CE} continues to increase because the base–collector depletion width increases, which means there will be fewer electrons recombining in the base.

9.9 Increase the bias voltage V_{CC}, and reduce the value of the collector resistor R_C.

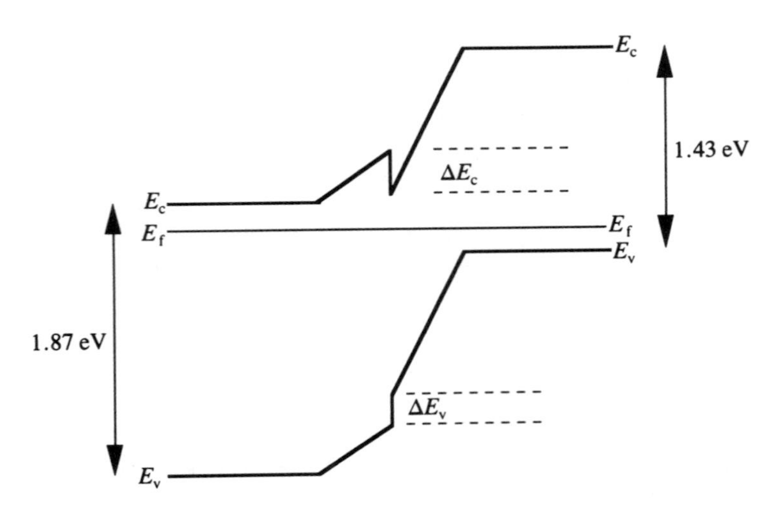

Figure SAQ 10.1 *The solution to SAQ 10.1.*

Chapter 10

10.1 See Figure SAQ 10.1.

10.2 1.6 eV is equivalent to a wavelength of 7.8 μm.

APPENDIX A

The electronic configurations of all the known elements

Z	Element		Electronic configuration
1	H	hydrogen	$1s^1$
2	He	helium	$1s^2$
3	Li	lithium	$1s^22s^1$
4	Be	beryllium	$1s^22s^2$
5	B	boron	$1s^22s^22p^1$
6	C	carbon	$1s^22s^22p^2$
7	N	nitrogen	$1s^22s^22p^3$
8	O	oxygen	$1s^22s^22p^4$
9	F	fluorine	$1s^22s^22p^5$
10	Ne	neon	$1s^22s^22p^6$
11	Na	sodium	$1s^22s^22p^63s^1$
12	Mg	magnesium	$1s^22s^22p^63s^2$
13	Al	aluminium	$1s^22s^22p^63s^23p^1$
14	Si	silicon	$1s^22s^22p^63s^23p^2$
15	P	phosphorus	$1s^22s^22p^63s^23p^3$
16	S	sulphur	$1s^22s^22p^63s^23p^4$
17	Cl	chlorine	$1s^22s^22p^63s^23p^5$
18	Ar	argon	$1s^22s^22p^63s^23p^6$
19	K	potassium	$1s^22s^22p^63s^23p^64s^1$
20	Ca	calcium	$1s^22s^22p^63s^23p^64s^2$

21	Sc	scandium	$1s^22s^22p^63s^23p^63d^14s^2$
22	Ti	titanium	$1s^22s^22p^63s^23p^63d^24s^2$
23	V	vanadium	$1s^22s^22p^63s^23p^63d^34s^2$
24	Cr	chromium	$1s^22s^22p^63s^23p^63d^54s^1$
25	Mn	manganese	$1s^22s^22p^63s^23p^63d^54s^2$
26	Fe	iron	$1s^22s^22p^63s^23p^63d^64s^2$
27	Co	cobalt	$1s^22s^22p^63s^23p^63d^74s^2$
28	Ni	nickel	$1s^22s^22p^63s^23p^63d^84s^2$
29	Cu	copper	$1s^22s^22p^63s^23p^63d^{10}4s^1$
30	Zn	zinc	$1s^22s^22p^63s^23p^63d^{10}4s^2$
31	Ga	gallium	$1s^22s^22p^63s^23p^63d^{10}4s^24p^1$
32	Ge	germanium	$1s^22s^22p^63s^23p^63d^{10}4s^24p^2$
33	As	arsenic	$1s^22s^22p^63s^23p^63d^{10}4s^24p^3$
34	Se	selenium	$1s^22s^22p^63s^23p^63d^{10}4s^24p^4$
35	Br	bromine	$1s^22s^22p^63s^23p^63d^{10}4s^24p^5$
36	Kr	krypton	$1s^22s^22p^63s^23p^63d^{10}4s^24p^6$
37	Rb	rubidium	$1s^22s^22p^63s^23p^63d^{10}4s^24p^65s^1$
38	Sr	strontium	$1s^22s^22p^63s^23p^63d^{10}4s^24p^65s^2$
39	Y	yttrium	$1s^22s^22p^63s^23p^63d^{10}4s^24p^64d^15s^2$
40	Zr	zirconium	$1s^22s^22p^63s^23p^63d^{10}4s^24p^64d^25s^2$
41	Nb	niobium	$1s^22s^22p^63s^23p^63d^{10}4s^24p^64d^45s^1$
42	Mo	molybdenum	$1s^22s^22p^63s^23p^63d^{10}4s^24p^64d^55s^1$
43	Tc	technetium	$1s^22s^22p^63s^23p^63d^{10}4s^24p^64d^65s^1$
44	Ru	ruthenium	$1s^22s^22p^63s^23p^63d^{10}4s^24p^64d^75s^1$
45	Rh	rhodium	$1s^22s^22p^63s^23p^63d^{10}4s^24p^64d^85s^1$
46	Pd	palladium	$1s^22s^22p^63s^23p^63d^{10}4s^24p^64d^{10}$

47	Ag	silver	$1s^22s^22p^63s^23p^63d^{10}4s^24p^64d^{10}5s^1$
48	Cd	cadmium	$1s^22s^22p^63s^23p^63d^{10}4s^24p^64d^{10}5s^2$
49	In	indium	$1s^22s^22p^63s^23p^63d^{10}4s^24p^64d^{10}5s^25p^1$
50	Sn	tin	$1s^22s^22p^63s^23p^63d^{10}4s^24p^64d^{10}5s^25p^2$
51	Sb	antimony	$1s^22s^22p^63s^23p^63d^{10}4s^24p^64d^{10}5s^25p^3$
52	Te	tellurium	$1s^22s^22p^63s^23p^63d^{10}4s^24p^64d^{10}5s^25p^4$
53	I	iodine	$1s^22s^22p^63s^23p^63d^{10}4s^24p^64d^{10}5s^25p^5$
54	Xe	xenon	$1s^22s^22p^63s^23p^63d^{10}4s^24p^64d^{10}5s^25p^6$
55	Cs	caesium	$1s^22s^22p^63s^23p^63d^{10}4s^24p^64d^{10}5s^25p^66s^1$
56	Ba	barium	$1s^22s^22p^63s^23p^63d^{10}4s^24p^64d^{10}5s^25p^66s^2$
57	La	lanthanum	$1s^22s^22p^63s^23p^63d^{10}4s^24p^64d^{10}5s^25p^65d^16s^2$
58	Ce	cerium	$1s^22s^22p^63s^23p^63d^{10}4s^24p^64d^{10}4f^25s^25p^66s^2$
59	Pr	praseodymium	$1s^22s^22p^63s^23p^63d^{10}4s^24p^64d^{10}4f^35s^25p^66s^2$
60	Nd	neodymium	$1s^22s^22p^63s^23p^63d^{10}4s^24p^64d^{10}4f^45s^25p^66s^2$
61	Pm	promethium	$1s^22s^22p^63s^23p^63d^{10}4s^24p^64d^{10}4f^55s^25p^66s^2$
62	Sm	samarium	$1s^22s^22p^63s^23p^63d^{10}4s^24p^64d^{10}4f^65s^25p^66s^2$
63	Eu	europium	$1s^22s^22p^63s^23p^63d^{10}4s^24p^64d^{10}4f^75s^25p^66s^2$
64	Gd	gadolinium	$1s^22s^22p^63s^23p^63d^{10}4s^24p^64d^{10}4f^75s^25p^65d^16s^2$
65	Tb	terbium	$1s^22s^22p^63s^23p^63d^{10}4s^24p^64d^{10}4f^95s^25p^66s^2$
66	Dy	dysprosium	$1s^22s^22p^63s^23p^63d^{10}4s^24p^64d^{10}4f^{10}5s^25p^66s^2$
67	Ho	holmium	$1s^22s^22p^63s^23p^63d^{10}4s^24p^64d^{10}4f^{11}5s^25p^66s^2$
68	Er	erbium	$1s^22s^22p^63s^23p^63d^{10}4s^24p^64d^{10}4f^{12}5s^25p^66s^2$
69	Tm	thulium	$1s^22s^22p^63s^23p^63d^{10}4s^24p^64d^{10}4f^{13}5s^25p^66s^2$
70	Yb	ytterbium	$1s^22s^22p^63s^23p^63d^{10}4s^24p^64d^{10}4f^{14}5s^25p^66s^2$
71	Lu	lutetium	$1s^22s^22p^63s^23p^63d^{10}4s^24p^64d^{10}4f^{14}5s^25p^65d^16s^2$
72	Hf	hafnium	$1s^22s^22p^63s^23p^63d^{10}4s^24p^64d^{10}4f^{14}5s^25p^65d^26s^2$

73	Ta	tantalum	$1s^2 2s^2 2p^6 3s^2 3p^6 3d^{10} 4s^2 4p^6 4d^{10} 4f^{14} 5s^2 5p^6 5d^3 6s^2$
74	W	tungsten	$1s^2 2s^2 2p^6 3s^2 3p^6 3d^{10} 4s^2 4p^6 4d^{10} 4f^{14} 5s^2 5p^6 5d^4 6s^2$
75	Re	rhenium	$1s^2 2s^2 2p^6 3s^2 3p^6 3d^{10} 4s^2 4p^6 4d^{10} 4f^{14} 5s^2 5p^6 5d^5 6s^2$
76	Os	osmium	$1s^2 2s^2 2p^6 3s^2 3p^6 3d^{10} 4s^2 4p^6 4d^{10} 4f^{14} 5s^2 5p^6 5d^6 6s^2$
77	Ir	iridium	$1s^2 2s^2 2p^6 3s^2 3p^6 3d^{10} 4s^2 4p^6 4d^{10} 4f^{14} 5s^2 5p^6 5d^9$
78	Pt	platinum	$1s^2 2s^2 2p^6 3s^2 3p^6 3d^{10} 4s^2 4p^6 4d^{10} 4f^{14} 5s^2 5p^6 5d^9 6s^1$
79	Au	gold	$1s^2 2s^2 2p^6 3s^2 3p^6 3d^{10} 4s^2 4p^6 4d^{10} 4f^{14} 5s^2 5p^6 5d^{10} 6s^1$
80	Hg	mercury	$1s^2 2s^2 2p^6 3s^2 3p^6 3d^{10} 4s^2 4p^6 4d^{10} 4f^{14} 5s^2 5p^6 5d^{10} 6s^2$
81	Tl	thallium	$1s^2 2s^2 2p^6 3s^2 3p^6 3d^{10} 4s^2 4p^6 4d^{10} 4f^{14} 5s^2 5p^6 5d^{10} 6s^2 6p^1$
82	Pb	lead	$1s^2 2s^2 2p^6 3s^2 3p^6 3d^{10} 4s^2 4p^6 4d^{10} 4f^{14} 5s^2 5p^6 5d^{10} 6s^2 6p^2$
83	Bi	bismuth	$1s^2 2s^2 2p^6 3s^2 3p^6 3d^{10} 4s^2 4p^6 4d^{10} 4f^{14} 5s^2 5p^6 5d^{10} 6s^2 6p^3$
84	Po	polonium	$1s^2 2s^2 2p^6 3s^2 3p^6 3d^{10} 4s^2 4p^6 4d^{10} 4f^{14} 5s^2 5p^6 5d^{10} 6s^2 6p^4$
85	At	astatine	$1s^2 2s^2 2p^6 3s^2 3p^6 3d^{10} 4s^2 4p^6 4d^{10} 4f^{14} 5s^2 5p^6 5d^{10} 6s^2 6p^5$
86	Rn	radon	$1s^2 2s^2 2p^6 3s^2 3p^6 3d^{10} 4s^2 4p^6 4d^{10} 4f^{14} 5s^2 5p^6 5d^{10} 6s^2 6p^6$
87	Fr	francium	$1s^2 2s^2 2p^6 3s^2 3p^6 3d^{10} 4s^2 4p^6 4d^{10} 4f^{14} 5s^2 5p^6 5d^{10} 6s^2 6p^6 7s^1$
88	Ra	radium	$1s^2 2s^2 2p^6 3s^2 3p^6 3d^{10} 4s^2 4p^6 4d^{10} 4f^{14} 5s^2 5p^6 5d^{10} 6s^2 6p^6 7s^2$
89	Ac	actinium	$1s^2 2s^2 2p^6 3s^2 3p^6 3d^{10} 4s^2 4p^6 4d^{10} 4f^{14} 5s^2 5p^6 5d^{10} 6s^2 6p^6 6d^1 7s^2$
90	Th	thorium	$1s^2 2s^2 2p^6 3s^2 3p^6 3d^{10} 4s^2 4p^6 4d^{10} 4f^{14} 5s^2 5p^6 5d^{10} 6s^2 6p^6 6d^2 7s^2$
91	Pa	protoactinium	$1s^2 2s^2 2p^6 3s^2 3p^6 3d^{10} 4s^2 4p^6 4d^{10} 4f^{14} 5s^2 5p^6 5d^{10} 5f^2 6s^2 6p^6 6d^1 7s^2$
92	U	uranium	$1s^2 2s^2 2p^6 3s^2 3p^6 3d^{10} 4s^2 4p^6 4d^{10} 4f^{14} 5s^2 5p^6 5d^{10} 5f^3 6s^2 6p^6 6d^1 7s^2$
93	Np	neptunium	$1s^2 2s^2 2p^6 3s^2 3p^6 3d^{10} 4s^2 4p^6 4d^{10} 4f^{14} 5s^2 5p^6 5d^{10} 5f^4 6s^2 6p^6 6d^1 7s^2$
94	Pu	plutonium	$1s^2 2s^2 2p^6 3s^2 3p^6 3d^{10} 4s^2 4p^6 4d^{10} 4f^{14} 5s^2 5p^6 5d^{10} 5f^6 6s^2 6p^6 7s^2$
95	Am	americium	$1s^2 2s^2 2p^6 3s^2 3p^6 3d^{10} 4s^2 4p^6 4d^{10} 4f^{14} 5s^2 5p^6 5d^{10} 5f^7 6s^2 6p^6 7s^2$
96	Cm	curium	$1s^2 2s^2 2p^6 3s^2 3p^6 3d^{10} 4s^2 4p^6 4d^{10} 4f^{14} 5s^2 5p^6 5d^{10} 5f^7 6s^2 6p^6 6d^1 7s^2$
97	Bk	berkelium	$1s^2 2s^2 2p^6 3s^2 3p^6 3d^{10} 4s^2 4p^6 4d^{10} 4f^{14} 5s^2 5p^6 5d^{10} 5f^8 6s^2 6p^6 6d^1 7s^2$
98	Cf	californium	$1s^2 2s^2 2p^6 3s^2 3p^6 3d^{10} 4s^2 4p^6 4d^{10} 4f^{14} 5s^2 5p^6 5d^{10} 5f^{10} 6s^2 6p^6 7s^2$

99	Es	einsteinium	$1s^22s^22p^63s^23p^63d^{10}4s^24p^64d^{10}4f^{14}5s^25p^65d^{10}5f^{11}6s^26p^67s^2$
100	Fm	fermium	$1s^22s^22p^63s^23p^63d^{10}4s^24p^64d^{10}4f^{14}5s^25p^65d^{10}5f^{12}6s^26p^67s^2$
101	Md	mendelevium	$1s^22s^22p^63s^23p^63d^{10}4s^24p^64d^{10}4f^{14}5s^25p^65d^{10}5f^{13}6s^26p^67s^2$
102	No	nobelium	$1s^22s^22p^63s^23p^63d^{10}4s^24p^64d^{10}4f^{14}5s^25p^65d^{10}5f^{14}6s^26p^67s^2$
103	Lr	lawrencium	$1s^22s^22p^63s^23p^63d^{10}4s^24p^64d^{10}4f^{14}5s^25p^65d^{10}5f^{14}6s^26p^66d^17s^2$
104	Rf [1]	rutherfordium	$1s^22s^22p^63s^23p^63d^{10}4s^24p^64d^{10}4f^{14}5s^25p^65d^{10}5f^{14}6s^26p^66d^27s^2$
105	Ha [1]	hahnium	$1s^22s^22p^63s^23p^63d^{10}4s^24p^64d^{10}4f^{14}5s^25p^65d^{10}5f^{14}6s^26p^66d^37s^2$
106	Sg [1]	seaborgium	$1s^22s^22p^63s^23p^63d^{10}4s^24p^64d^{10}4f^{14}5s^25p^65d^{10}5f^{14}6s^26p^66d^47s^2$
107	Ns [1]	nielsbohrium	$1s^22s^22p^63s^23p^63d^{10}4s^24p^64d^{10}4f^{14}5s^25p^65d^{10}5f^{14}6s^26p^66d^57s^2$

[1] Names adopted by the American Chemical Society but awaiting approval from the International Union of Pure and Applied Chemistry. The IUPAC's Commission on Nomenclature of Inorganic Chemistry has proposed dubnium Db, joliotium Jl, rutherfordium Rf and bohrium Bh respectively for these four elements. At the time of writing the result of the dispute between the ACS and the IUPAC is not yet known.

APPENDIX B

The p–n junction at equilibrium: proof that the Fermi level E_f is uniform throughout

We know that, for a p–n junction in thermal equilibrium, the electron current density J_n is the sum of the electron drift current density and the electron diffusion current density:

$$J_n = J_{n(\text{drift})} + J_{n(\text{diffusion})}$$

$$= e\mu_n n_o \mathcal{E} + eD_n \frac{dn}{dx} \tag{B1}$$

Likewise, we can write a similar expression for the hole current density J_p:

$$J_p = J_{p(\text{drift})} + J_{p(\text{diffusion})}$$

$$= e\mu_p p_o \mathcal{E} - eD_p \frac{dp}{dx} \tag{B2}$$

We also know that

$$J_n = 0$$

$$J_p = 0$$

at equilibrium.

What about the Fermi level at equilibrium? Figure B1 shows the band diagram for an abrupt p–n junction at equilibrium.

You can see in Figure B1 that $E_{fp} = E_{fn}$ at equilibrium. Proof of this follows.

The built-in electric field $\mathcal{E}$ can be written in terms of electron energy E:

$$\mathcal{E} = \frac{1}{e}\frac{dE}{dx} \tag{B3}$$

Remember energy E is in electron volts, so dE/e will be in V.

Across the depletion width W the built-in electric field $\mathcal{E}$ causes a force to be exerted on each electron entering the depletion region of

$$F_{\text{elec}} = -e\mathcal{E} \tag{B4}$$

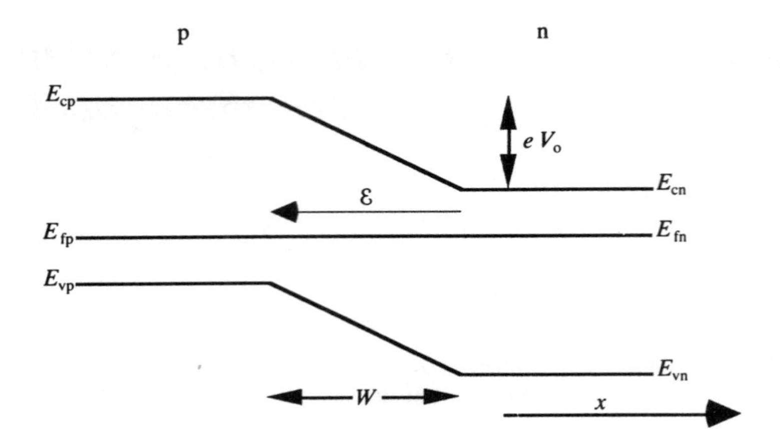

Figure B1 *Abrupt p–n junction at equilibrium.*

Hence

$$F_{elec} = -\frac{dE}{dx} \tag{B5}$$

F_{elec} therefore equals the slope of electron energy E versus distance x across the depletion region. We could use any electron energy in these equations; for instance, we could replace E by E_c:

$$\mathcal{E} = \frac{1}{e}\frac{dE_c}{dx}$$

or we could replace E by E_v:

$$\mathcal{E} = \frac{1}{e}\frac{dE_v}{dx}$$

or by E_i:

$$\mathcal{E} = \frac{1}{e}\frac{dE_i}{dx}$$

because E_c, E_v and E_i all have the same slope.
Hence our earlier expression for J_p (equation (B2)) can be rewritten:

$$J_p = e\mu_p p_o \frac{1}{e}\frac{dE_i}{dx} - eD_p\frac{dp}{dx} \tag{B6}$$

We can rewrite the right-hand side using the Einstein relationship

$$D_p = \frac{kT\mu_p}{e} \tag{B7}$$

Hence

$$J_p = e\mu_p p_o \frac{1}{e}\frac{dE_i}{dx} - kT\mu_p \frac{dp}{dx} \tag{B8}$$
$$= 0$$

Remember equilibrium conditions, therefore there must be no net current.
However,

$$p_o = n_i \exp[(E_i - E_f)/kT] \tag{B9}$$

and

$$\frac{dp_o}{dx} = \frac{d}{dx}\{n_i \exp[(E_i - E_f)/kT]\} \tag{B10}$$

$$= \frac{n_i \exp[(E_i - E_f)/kT]}{kT} \frac{d}{dx}(E_i - E_f)$$

$$= \frac{p_o}{kT}\left[\frac{dE_i}{dx} - \frac{dE_f}{dx}\right] \tag{B11}$$

Now consider equation (B8). We can substitute equation (B11) into equation (B8) to get:

$$J_p = e\mu_p \frac{1}{e}\frac{dE_i}{dx} - kT\mu_p \frac{p_o}{kT}\left[\frac{dE_i}{dx} - \frac{dE_f}{dx}\right]$$

$$= \mu_p p_o \frac{dE_i}{dx} - \mu_p p_o \frac{dE_i}{dx} + \mu_p p_o \frac{dE_f}{dx}$$

$$= \mu_p p_o \frac{dE_f}{dx} = 0 \tag{B12}$$

But

$$\mu_p p_o \neq 0$$

therefore

$$\frac{dE_f}{dx} = 0$$

This shows that the slope in dE_f across the junction must be zero. We can derive a similar expression for electrons:

$$J_n = \mu_n n_o \frac{dE_f}{dx} = 0 \tag{B13}$$

We have thus proved that, at equilibrium, the Fermi level E_f must be uniform (i.e. independent of x) throughout the p–n junction.

Index